Applied Geological
Micropalaeontology

P. K. Kathal
Dept. of Applied Geology
Dr. Harisingh Gour University
(A Central University)
Sagar-470003 (M.P.), India
pkkathal@dhsgsu.ac.in pkkathal@rediffmail.com

SCIENTIFIC
PUBLISHERS (INDIA)

Published by:
Scientific Publishers
5A New Pali Road, P.O. Box 91
Jodhpur 342 001 (India)

E-mail: info@scientificpub.com
Website: www.scientificpub.com

ISBN: 978-81-7233-753-7 (HB)
978-81-7233-754-5 (PB)

Typeset by Rajesh Ojha
Printed in India

Dedicated to
my revered teacher
Professor S. N. Bhalla

WADIA INSTITUTE OF HIMALAYAN GEOLOGY
(An Autonomous Institution of Deptt. of Science & Technology, Govt, of India)
33, General Mahadeo Singh Road
Dehra Dun - 248001 (Uttarakhand)

Prof. Anil K. Gupta, RNASC., FA.SC
Director

Foreword

This book entitled "Applied Geological Micropaleontology' is primarily intended as a basic text for graduate and postgraduate students studying Micropalconotlogy. Because of wide applications of microorganisms in living or fossil forms, this branch is normally studied as a part of geological curriculum. Since this discipline is concerned with the remains of once-living microorganisms it is also a part of biological subjects. As biological background of students reading geology may be limited, the author has dwelt upon adequate aspects of microfossils to make their taxonomy interpretative.

Within the past seven decades the development of micropaleontology in terms of both information and concept has been almost explosive. The form and structure of microfossils is now known in great details, particularly due to scanning electron microscope and availability of microfossils recovered from marine cores under programs like the Deep Sea Drilling Project (DSDP), Ocean Drilling Project (ODP), etc. This has led to publication of enormous data on one hand and gradual refinement of taxonomy on the other; arousing a need to synthesize the existing data in a simplified text-book format.

Among microfossils, foraminifera, ostracoda, calcareous nannoplanktons, diatoms, conodonts have vast deal of applications and have been given due coverage in the book. Breaking the traditional barriers of descriptive taxonomy and biostratigraphy the author has made an effort to make the taxonomy interpretative so that the existing ecological data could be transformed in the 'rich-text' enabling the students to understand the trend of the present-day exploration techniques.

The author of the book Professor P. K. Kathal during the last 30 years of teaching, research and interaction with students realized a need to fill the widening-gap between the applied-researches on the various groups of microfossils and the existing text-books and has contributed this book, which I am sure, will motivate the students towards the fascinating world of microfossils.

(Professor Anil K. Gupta)
FASc, FNASc
Director, Wadia Institute of Himalayan Geology
Dehra Dun - 248 001, India

Dehra Dun 15
August 2011

Preface

Microfossils have wide applications in the field of Geology, Zoology, Botany, and Oceanography. After contributing a book entitled 'Microfossils and their applications' in 1998, I felt that the gap between the researches in the field of Micropaleontology and the existing text-books is widening tremendously.

In the last 25 years Micropaleontology has emerged as an advanced branch of earth sciences progressing at a rather fast-pace than expected, owing to its interpretative taxonomy that helped us solving enduring problems in the fields of Geology and Oceanography especially the higher resolution biostratigraphy, paleoenvironment, basin evolution for petroleum exploration, climatology etc.

The present book entitled 'Applied Geological Micropaleontology' incorporates the text on the various microfossil groups. It is endeavoured to incorporate the existing knowledge in a systematic, comprehensive and expressive manner enabling the readers including those with mathematical background to choose Micropaleontology as their careers.

The book consists of 7 chapters, plates, appendix, bibliography, general index and systematic index. Chapter 1 is intended to provide the background and fundamental information required to delve into the subject through further chapters; Chapter 2 deals with the various techniques employed in the collection of samples, separation of microfossils, illustrations, micrography, intraspecific variation and migration studies, statistical and computational studies; Chapter 3 is devoted to the various aspects of foraminifera, their classifications, ecology, applications, geological distribution etc.; Chapter 4 is devoted to Diatoms, their classifications, ecology, applications; Chapter 5 covers aspects on the Calcareous Nannoplanktons, their classifications, ecology, applications, geological distribution etc.; Chapter 6 is devoted to Ostracoda, their morphology, taxonomy, ecology, geological applications and geological distribution; and Chapter 7 the various aspects of Conodonta, techniques of their studies, biology, composition and ecology. Besides, each most of the chapters are supplemented with illustrative Glossary at the end.

To summarise this book provides a logically arranged text on microfossils that forms part of the syllabus at postgraduate level in the institution offering Geology, Applied Geology and Petroleum Geology.

I would welcome suggestions from readers on the shortcomings left in the text.

21st July 2011

Sagar

P. K. Kathal

pkkathal@dhsgsu.ac.in

pkkatha@rediffmail.com

Acknowledgement

First of all I am thankful to Professor A. K. Gupta, F.N.A.Sc., F.A.Sc., Director Wadia Institute of Himalayan Geology, Dehradun for going through the manuscript of the book and writing its 'Foreword'.

I am thankful to Prof. Y. Matoba, Akita University, Japan, Prof. S. Hasegawa, Kumamoto University, Japan, Prof. M. Kucera, Prof. C. Hembleben, Dr. H. Schultz, Dr. P. Heinze, Tuebingen University, Tuebingen, Germany, for productive discussions on the latest trend of researches in the field of Micropaleontology.

I am thankful to the Prof. N. S. Gajbhiye, Vice-Chancellor (F.N.A. Sc.), and my teachers Prof. B. Das, Prof. M. V. Durge, Prof. S. N. Pandey, Prof. P. O. Alexander, Dr. D. P. Khare, Prof. L. P. Chourasia, Dr. Harisingh Gour University, Sagar for their moral support.

I sincerely thank Prof. A. K. Jauhri, Lucknow University for his help in developing text on larger foraminifera; Prof. K. L. Shrivastava, Jodhpur University, Prof. K. N. Gaur, Aligarh, Prof. P. Dev, Ujjain University, Prof. Abu Talib, Aligarh Muslim University, Aligarh, Dr. Rajiv Nigam, National Institute of Oceanography, Goa; Prof. P. K. Saraswati, Indian Institute of Technology, Powai, for fruitful discussions; and Prof. A. K. Shandilya, Ajmer University, Ajmer and Prof. Arun Kumar, Manipur University, Imphal, for moral support.

I thank Prof. A. K. Shandilya, Prof. R. K. Trivedi, Dr. S. K. Gupta, Dr. R. K. Rawat, Dr. S. H. Adil, Dr. H. Thomas, Dr. K. K. Prajapati and Dr. M. K. Purohit, the ministerial-staff and students of the Department of Applied Geology and the Centre of Advanced Study in Geology, Sagar and my friends Dr. R. K. Baronia, Chhindwara, Shri P. P. Singh, Dr. J. K. Mishra, Prof. S. J. Singh, Sagar for their help.

I am also thanking Dr. V. K. Singh, Ku. Jemini Khateek for their assistance; and my aunt Smt. Dropadi Kathal, wife Geeta and daughter Priyanshi for over-all support during the completion of this project.

21st July 2011 **P. K. Kathal**

Sagar

Contents

CHAPTER 1

Introduction to Microfossils

Contents

1.1 INTRODUCTION

Palaeontology, coined by de Blanville and Von Wolderheim (1834) is a discourse on study of fossils (fossil, Latin *fodere,* dug-up), the remains of organisms that by certain process, have become a part of rocks, they are found embedded in and may be considered as 'evidences of life' of pre-historic times. It has gradually replaced the term 'orycotology'.

Palaeontology (Paleontology) has evolved through Zoology, which is a discourse on the living animals and may be subdivided into: (a) Neontology, related to the 'existing' life and (b) Palaeontology or Paleobiology, to the 'ancient' or extinct life (Fig. 1).

Palaeontology has gained the status of an essential tool for stratigraphers working on various aspects viz. stratigraphic correlation, strata-identification, high-resolution stratigraphy, event stratigraphy, basin-evolution, establishment of sequences and determination of paleoenvionment, paleoecology and paleogeography. From biological points of view it yields important evidences for the study of evolution and adoption of organisms to different environments.

Microfossils require special methods of collection, separation, illustration and identification and can only be studied under microscope. Therefore, Micropalaeontology was introduced as a separate branch of Geology in 1863 by Ford to study fossils of small invertebrates (< 0.1 mm, diameter/length) e.g., foraminifera, diatom, radiolaria, ostracoda, nannoplankton, brayozoa etc. or parts of macro-invertebrates whose study require use of a microscope throughout. It incorporates study of organisms of the geological past, their structures and genetic relationships in time and space.

Microfossils have a vast application in various geological areas especially in the biostratigraphy and petroleum exploration owing to their great abundance in time and space, wide morphological variability, and usefulness in paleoenvironmental interpretations. In the Indian context, their application in the study of paleomonsoonal precipitation has gained momentum in the last three decades. The types of microfossil groups, their composition and applications have been enumerated in Table 1.1.

The Scanning Electron Microscopy (SEM), however, revolutionised taxonomy and applications of microfossils in multifaceted ways.

1.2 HISTORY OF DEVELOPMENT OF MICROPALAEONTOLOGY

The modern history of Micropalaeontology is no older than seven decades. However, the first mention of fossil in the literature dates back to the Greek natural philosopher Herodotus in the 5th century when he noticed coin shaped Foraminifera (*Nummulites)* embedded in limestone blocks used in the Egyptian pyramids and described them as 'stone lentis' or 'coin fossil'. In middle of the 16th century Agricola and Gesner and in the early 17th century Scheuchzer recognised them as 'stone fossils'. The advent of microscope in 1660 by Leeuwenhoek accelerated the pace of study of microfossils of Tertiary sediments. Beccarius (1731) gave systematic description of micro-foraminifera. In the middle of the 17th Century, Plancis published a monograph on Recent foraminifera of sediments.

Table 1.1 Important groups of microfossils, their compositions and applications

Composition		Group	Application
Organic-walled (palynomorphs)		**Dinoflagellate** cysts (calc./siliceous) Marine/fresh waters; Cyst 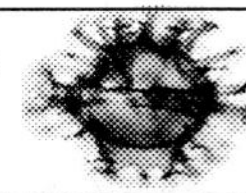	Abundant, migrate distantly through wind/water; uses: paleoecology & paleoenvironment; studies; distance to land in marine rocks; markers in freshwaters & deltaic rocks; correlation of continental & near-shore marine in the Devonian or younger deposits .
		Pollen & spores Terrestrial/marine  Spores Pollen	Biozonation of marine Mesozoic/Cenozoic rocks with reworked cysts & spores: provenance & direction of transport; based on kinds of cysts; near- & off-shore Quaternary stratigraphy; probable climate-controls, correlation & interpretation of marine strata.
Mineral Walled	**Calcareous** (Some arenaceous)	**Calcareous** **Arenaceous** **Foraminifera**  Shallow & marine 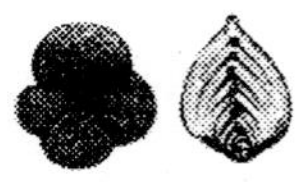 Planktic Benthic	High-resolution biostratigraphy; fusulinids in bio-zonning of U. Carboniferous-Permian; Paleotemp.; interpretative taxonomy; O^{18}/O^{16} ratio for paleotemp. of surface (planktic) & bottom (benthic) waters; planktics in biostratigraphy; offshore deep-water oil-exploration; marker in Cre.- Rec.; Calcareous species show temp. above 12^0 C, shallow depth / low latitudes; arenacoeus species, cold , very deep waters & high latitudes.
		Calcareous nannoplankton Marine 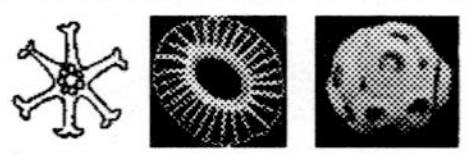 Discoaster Coccolith Coccosp.	High resolution biostratigraphy of Mesozoic & Cenozoic rocks; surface water paleotemp.; paleoecology; Miocene-water-masses shift; glacial/interglacial zone (Pleistocene); O^{18}/O^{16} ratio to study temp. & chemistry of surface, column, bottom waters; & rate of sedimentation. (Coccosp.= Coccosphere)
	Siliceous	**Radiolaria** Marine, deep water 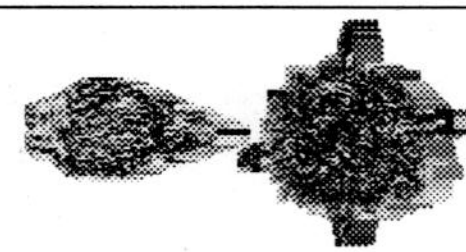	Marine Mesozoic biostratigraphy; high-latitude, deep sea-sediment study because of no effect of desolation; interpretative taxonomy; evolution correlates well with magnetic reversal events of last 3 Ma phenotypic variations in carapace morphology helps in delineating the 'outer bay' and 'inner bay' facies in estuarine deposits.
		Diatom Mainly marine Planktonic 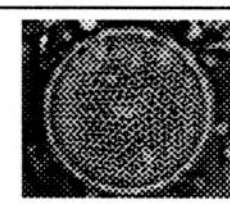	Bio-zones (Cretaceous & Tertiary in freshwater rocks); warm-cool water bio-zonation; paleoecologic value (Pliocene-Pleistocene); silica tests used in O^{16} & O^{18} based temperature studies; fresh-water forms, in lake development history; effects of acidity in climate & effects of human-population.
	Phosphatic	**Conodonta** Extinct 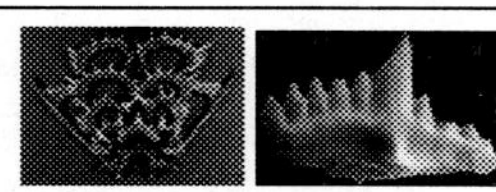 *Odontogriphus* & its element	Zoning of Ordovician, Permian-Triassic marine rocks; sea-floor spreading; change in their color shows influence of temperature & duration of burial; lamellae indicates astronomical days, nights, tides; paleo-salinity; rate of sedimentation; degree of reducing environment; current Index; conversion of organic matter; chances of nearby oil pools.
	Chitino-calcareous	**Ostracoda** Marine/ freshwater 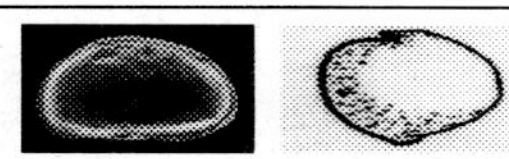 Benthic Planktic	Paleo-salinity/shorelines, varied morphology for dating, biostratigraphy and paleoenvironment; oil exploration using ratio of open/closed carapaces in rate of sedimentation; degree of pyritization & degree of the reducing environment; degree of deformation for study of compaction of rock during palengenesis.

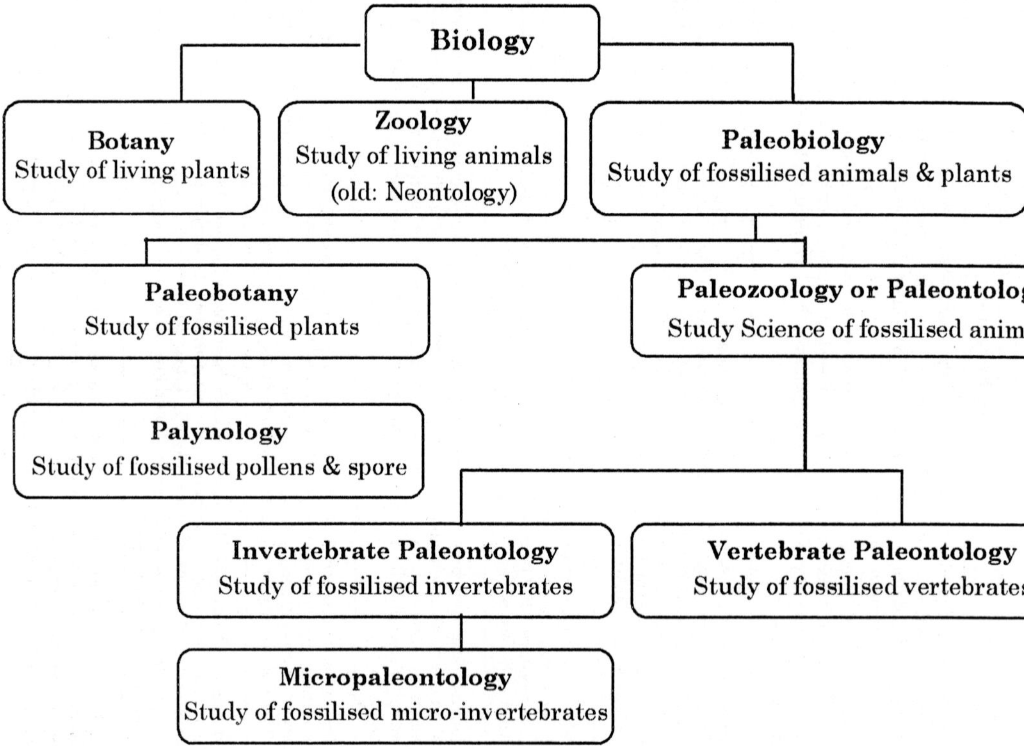

Figure 1.1 Micropalaeontology and allied branches of Biology

Linne (1758) published his monumental work, *Systema Naturae* (classification of the animals and plants), which initiated the nomenclature of 'genus' and 'species'.

Although the famous French zoologist Lamarck, in 'Zoological Treatise' referred to the foraminifera as coral or cephalopod but most of the generic and specific names proposed by him are still valid. d'Orbigny (1802-1857), who is considered as the 'Father of Modern Micropalaeontology' published numerous papers describing and sketching several genus and species of foraminifera. He also grouped them as juvenile Cephalopods. Dujardin (1835) was the first to recognise them as protozoan.

Reuss and Terquem in mid-eighteen century studied Foraminifera and Ostracoda. The former also proposed his classification on foraminifera. The famous German micropalaeontologist C. G. Ehrenberg (1854) published his monumental work entitled *Mikrobiologie* that included Foraminifera, Ostracoda and diano-flagellates, and appreciated their roles in rock building processes.

The works of Carpenter, Parker and Jones (1862) and Lister during 1894-1903 also included very important aspects of foraminifera such as discovery of 'micro-' and 'megalospheric' tests (dimorphism), ontogenetic stages (growth stages of an individual organism) and phylogenitic stages (evolution stages of species through geological time). The first commercial use of microfossils was made in 1877 by Howchin when the middle

Miocene 'well-hole samples' of the Vienna Basin, Austria was studied using their foraminiferal contents. Chapman and Parr (1900-1920) and Schuchert from years 1890-1924 determined the ages of various rocks using microfossils, from all over the world.

The Catalogue of Foraminifera was published (Ellis and Massina, 1940, with supplements) covering about 1550 genera and over 17000 species. Subsequently, the Catalogues of Ostracoda and nannoplanktons were also published

J. A. Udden highlighted the importance of microfossils in developing high-resolution biostratigraphy, biostratigraphic-correlation and determination of age. He also convinced the then major oil-exploration companies of the importance of the study of microfossils in their work.

J. J. Galloway and J. A. Cushman published a series of significant papers on the morphology, ontogeny and classification of foraminifera in the early-middle-nineteenth century. An important milestone in the field of researches on foraminifera, 'Cushman Foundation of Foraminiferal Research' was established in 1924 in U.S.A. that started publication of research journal *'Contributions of the Cushman Laboratory'* (presently: *Journal of Foraminiferal Research*).

The 'Society of Economic Palaeontologists and Mineralogists' (SPEM) was founded jointly with the American Association of Petroleum Geologists (AAPG) to initiate publication of the *Journal of Palaeontology* in 1926. In 1927 Cushman published a book entitled "Foraminifera" and proposed the first comprehensive classification of foraminifera. Professor Galloway's "Manual of Foraminifera" appeared in 1933 with a different approach of classification of Foraminifera, which gained a great popularity amongst the micropaleontologists.

As stated earlier the advent of Scanning Electron Microscope (SEM) enabled work on the skeletal composition of microfossils up to a magnification of 10,00,000 times that helped reclassifying microfossils. A. R. Loeblich and Helen Tappan (1964, 1984 and 1988) classified foraminifera after removing the invalid texa for the taxonomy and by proposing new chains of phylogenetic relation.

H. N. Coryell and S. Fields (1937) highlighted the use of Ostracoda as an alternative of Foraminifera in certain environments like- very low salinity oceanic and fresh water sedimentary environments that remain almost devoid of foraminifera. Van Morkhovan recognised the importance of indentifying Palaeozoic and post Palaeozoic taxa of ostracoda separately that helped significantly in preparations of biostratigraphy zones.

Nannoplanktons were first recognised by Ehrenberg as very small, button-shaped discs as 'agaric mineral' or 'calcareous calcite discs and later regarded as 'morpholiths'. Huxley (1858-1968) referred to them as 'coccoliths'. Sorby (1861) and Wallich (1877) recognised that they were parts of larger spherical objects 'coccosphere'. Under the famous H. M. S. Challenger expedition, Murray and Renard highlighted their importance as member of the marine food-chain. Schiller (1930), Bernard (1953 to 1974) and G. Deflandre (1939 to 1954) studied the extinct-coccolithophores of their time. Kamptner (1931 to 1954) coined the term 'nanno-paleontology'. Bramlette and Riedel (1954) used them for world-wide correlation of Mesozoic and Tertiary pelagic sequences. In late 1960's, their vertical geological ranges helped worldwide biostratigraphic zonation (Hay, Bramlette and others) in tropical-subtropical and temperate regions of Tertiary strata. Leoblich and Tappan (1966 to 1973) published annoted index and bibliography. Haq and Bramlatte established practical biostratigraphic zonation of Tertiary strata. Hanjo and Okada (1976) showed control of surface water temperature on their distribution Haq and Van Couvering (1977) used them for correction of Pliocene/ Pleistocene boundary.

A Russian, C. H. Pander coined the name 'conodont' in 1856 with an assumption that they belonged to a soft bodied animal. Fay (1952) linked them to 'fish-like chordate'. Jones (1956) established their affinity to worm-like invertebrate. Ulrich and Bassler (1926) described many new species and distinguished their biostratigraphic usefulness. Schmidt (1934) and Scott (1934) found a set of conodont-elements and opined separately that the individual elements were in life held, likened to mouth parts. Since 1960's they are being used in Palaeozoic and Triassic biostratigraphy. Lelton and Scott found a fossil-structures showing 'chordate affinity'. Discovery of *Odontogriphus* (Conway-Morris, 1976) showed that the animal had annular-rings (like earthworm). Landing (1977) showed that they were used as grasping mechanism. They are now believed to be part of the feeding apparatus of an extinct early worm-like vertebrate *Clydagnathus.*

Diatoms have been studied since the late eighteenth century, but the studies picked-up pace in the early nineteenth century with the emerging developments in microscopy. European workers namely, Cleve, Ehrenberg, Grunow, Schmidt and van Heurck produced hand-illustrated monographs on diatoms in the late nineteenth-century. In the early twentieth-century their fossils were recovered and studied by Hustedt (1927-66) who produced their first taxonomic and ecological account that proved to be a key reference today. Round *et al.* (1990) did a comprehensive work on diatoms.

The literature published on microorganisms included more than 5000 research papers on Foraminifera, and 1200 on Ostracoda, Conodonta and other microfossils.

The leading journals published in the field of Micropalaeontology are: *Journal of Foraminiferal Research*, USA; *Journal of Micropalaeontology*, UK; *Micropalaeontology*, USA; *Journal of American Association of Petroleum Geologists*, USA; *Revue de Paleobiologie*, Switzerland; *Paleogeography, Paleoclimatology, Paleoecology*; *Marine Micropalaeontology*; *Paleontological Research*, Japan; *Neus Jaharbuch Geologie and Paleontologie*, Germany; *Revista Esponola de Paleontologia*, Spain; *Journal of Palaeontological Society of India*; and *Bulletin of the Oil and Natural Gas Corporation* (ONGC), India.

The smallest unit of taxonomy of life is a 'species' that forms a group of closely associated organisms that not only possesses common distinctive features and ecological niche but also interbreeds and reproduces its characteristics in young. Ideally these species share common ecological niche. However, it has been a difficult task for the taxonomists to establish the free-breeding in micro-organisms particularly in the extinct ones, until and unless they are sufficiently plentiful to allow recognition of morphological and ecological discontinuities.

Classification of living organisms

LIVING ORGANISMS

1. Superkingdom: Prokaryota
(unicellular without nucleus)
- 1. Subkingdom :**Monera**
  - 1. Phylum: **Cyanophyta***
    (Blue-green algae)
  - 2. Phylum: **Schizomicophyta**

2. Superkingdom : Eukaryota
(unicellular, multicellular organisms, with nucleus)
- 2. Kingdom : **Protista***
  (unicellular, with nucleus)
  - 1. Subkingdom : **Protophyta**
    (unicellular, autotrophic organism, mainly lower algae)
    - 1. Phylum : **Euglenophyta**
    - 2. Phylum : **Chrysophyta***
      (unicellular, golden pigment, hyptonema: Nannoplanktons
    - 3. Phylum : **Pyrrophyta**
    - 4. Phylum : **Hyphochytridiomycota**
    - 5. Phylum : **Plasmodiophoromycota**
  - 2. Subkingdom : **Protozoa***
    (unicellular, heterotrophic organisms)
    - 1. Phylum : **Sporozoa**
    - 2. Phylum : **Cindosporida**
    - 3. Phylum : **Zoomastigina**
    - 4. Phylum : **Sarcodina***
      (unicellular animals with test: Foraminifera, Radiolaria)
    - 5. Phylum : **Ciliophora**
- 3. Kingdom : **Fungi**
- 4. Kingdom : **Plantae***
- 5. Kingdom : **Animalia***

*** also occur as fossils**

Nutrition mode: Prokaryota: ingestion; Protophyata & Plantae: photosynthesis & Animali adsorption

Kingdom **Fungi**: Phylum **Oomycota, Chitridiomycota, Ascomycota, Basidiomaycota, Zygomycota, Myxomycota, Acrasiomycota & Labrinthulomycota.**

Kingdom **Plantae: Rhodophyta, Chlorophyta, Bryophyta, Tracheophyta & haeophyta.**

Kingdom **Animalia: Platyhelimenthis, Pogonophora, Trilobita, Chelicerata, Crustacea, Uniramia, Hexapoda, Enteropneusta, Chordata, Urochordata, Petrobranchia,** Echinodermata, Ectoprocta, Brachiopoda, Phoronidea, Annelida, **Mollusca, Aschelminthes & Porifera**

Figure 1.2 Classification of organisms (after: Whittakar, 1969).

In the next stage of hierarchy all the species are placed in a single 'genus'. Species of one genus should differ from the species of other genus with an acceptable morphological and ancestral gap. From the geological considerations, genera offer a very meagre value for stratigraphic correlation owing to their wide distribution in the time and space as compared to the species. Whereas they have great utility in the work related to paleoecological and paleogeographical studies. A group of genera constitutes a 'subfamily', or directly a 'family' in the order of hierarchy followed by 'superfamily', 'suborder' (or directly an 'order'), 'class', 'phyla', 'subkingdom', 'kingdom' and ultimately a 'superkingdom'.

Whittaker (1969) divided the world of organisms into two superkingdoms (one incorporating the prokaryotes and the other, the eukaryotes). The Eukaryota to which most of the microfossils belonged to were divided into 4 Kingdoms as shown Fig. 1.2.

Stratigraphic Micropalaeontology

In practice the fossil-based stratigraphy or biostratigraphy, of a fossiliferous succession can be classified into a biostratigraphic-column with the youngest rock on the top, supplemented by radioactive dating. The units are recognised and named after their fossil contents and others aspects like, the time of appearance, disappearance, maximum and minimum population of species/assemblages etc.

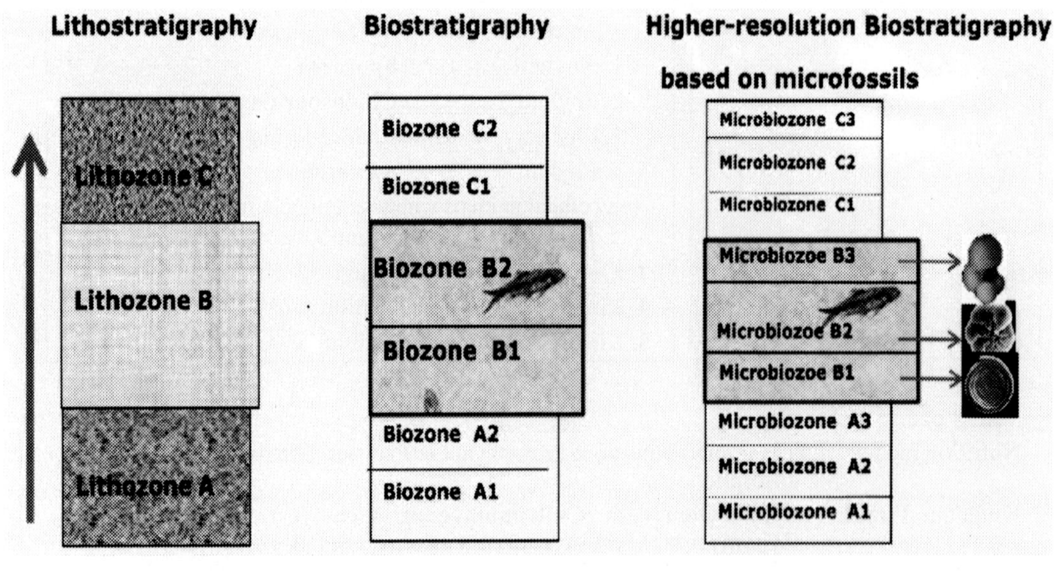

Figure 1.3 High-resolution time-slicing of a stratigraphic-column

Biozone, a fundamental biostratigraphic unit comprises rocks that are characterised by the occurrence of one or more specified fossils known as 'zone fossil(s)'. However, biozones may be divided into sub-biozone or microbiozones on the basis of microfossils to classify the succession at higher biostratigraphic resolution as shown in Fig. 1.3. Such studies help solving stratigraphic boundary problems at regional and global levels that has made possible world-wide correlation of stratigraphic units.

GLOSSARY OF SELECTED TERMS

Acme or peak zone. A body of strata in which the maximum abundance of a particular species is found and, which is less than the total range of the species.

Assemblage zone. The beds or a group of beds with a natural assemblage of fossils. Such zones are generally environmentally controlled and, therefore, used in local biostratigraphic correlation.

Autecology. The relationship of an individual organism and the environment in which it occurs.

Asexual reproduction. Multiplication not involving meiotic reduction division and fusion of gametes.

Bed. A subdivision of a formation or a member that has smaller thickness and lateral extent.

Benthic or benthonic. The organisms, which live on bottom or substratum of aquatic system. (**vagile, sessile benthos**)

Biocoenose. A group of organisms, which has adopted/inhabited an area of a given set of physico-chemical conditions.

Bioherms. The environment developed around a large coral reef.

Bionomical zones. The depth-zones of oceans viz. **littoral**, **neritic**, **bathyal** and **abyssal** zones in which the benthic forms are found distributed.

Biostromes. Flat and laminar-communities of reef-type animals that barely rise above the sea floor.

Biotope. The place of habitat of a biocoenose.

Biozone. The fundamental biostratigraphic unit that comprises those rocks that are characterised by the occurrence of one or more specified kinds of fossils called zone fossils.

Burrowing organisms. The **benthic** organisms that have a habit of burrowing.

Chemostratigraphy. Study of stratigraphic sequences on the basis of stable isotopes such as oxygen, carbon and strontium.

Chronostratigraphic units. The Eonothem, Erathem, System, Series, Stage are equivalent to the Aeon, Era, Period, Epoch, Age and Chron of the Geochronological units, respectively.

Chitin. A nitrogen containing polysaccharide.

Chronozone. The fundamental zone, which represents a specific time interval and used in chrono-stratigraphy in which a well-established and recognisable species was in existence.

Cyclostartigraphy. Study of short-period, high-frequency, sedimentary cycles in the stratigraphic-record, particularly by use of oxygen-isotopes data.

Derived fossils/recycled fossil. A group of fossils of certain older rock, re-deposited in the rock unit in which they are found now. Such fossils are useful as they indicate the upper age limit of the rock formations in which they occur now.

Diachronous fauna. A fauna where the species may not represent synchronity i.e. similar time-unit, owing to the possibility of their inward or outward migration from or to the environment of their occurrence.

Dimorphism. Organism occurring in two distinct (usually morphological) form

Diploid. A generation, having two of such chromosome in each **somatic** cell.

Disjunct species. The species whose occurrence in an environment is due to its migration from its original faunal-province.

Diversity of species. The number of species at a particular time and space.

Ecology. It primarily deals with the relationship of the organisms, to their environment in which they occur.

Endemic species. The species, which occur only in a particular faunal-province.

Epoch. A time-span during which the part of a **system** called **series** is deposited.

Euryhaline fauna/species. A group of marine organisms, which can also survive in reduced salinities e.g., **paralic environments** like estuaries, brackish water lakes etc.

Event stratigraphy. Correlation of sedimentary units on the basis of marker-beds or event-horizons.

Faunal province. Large geographical-regions of the sea or along the continental-selves within which the faunas at the specific, generic and even at the family level have a distinct identity.

Formation. The fundamental unit of the lithostratigraphic classification, which is a mappable litho-stratigraphic unit of an intermediate scale (1: 50,000).

Fossil of facies or facies fossil. Environment-controlled dissimilar assemblages of fossils (faunas), which represent different facies of rocks.

Gamete. A haploid germ-cell capable of fusing with the germ cell of the opposite sexuality to form a zygote e.g., spermatozoon and ovum.

Group. The broadest unit of lithostratigraphic-classification, comprises a thick-succession of rock formations spread over a large area.

High-resolution biostratigraphy. Study based on fine-scale stratigraphic-resolution using microfossils, nannofossils etc.

Interval zone. An interval between two distinct biostratigraphic-horizons.

Magnetostratigraphy. Deals with stratigraphic relationship on the basis of magnetic properties of sedimentary rocks and layered volcanic rocks.

Member. A part of the formation, which has some remarkable lithology or a characteristic fossil assemblages.

Nekton. Pelagic animals capable of swimming or making progress against natural water flow.

Ontogeny. The stages of development of an individual organism from a zygote to adult.

Orthogenesis. Unidirectional evolutionary trend.

Paleoclimate. Climate of a definite geological time unit.

Paleoecology. Deals with the relationship of fossil animals, to their environment, both as individuals (autecology) and in the faunal-communities (synecology) in which they occured in natural conditions at a particular geological time.

Paleoenvironment. Environment of the geological past.

Paralic environment. The marginal marine-environment where the salinity of water is less then the normal (> 30‰) e.g., estuary, bay, brackish-water lakes etc.

Pelagic organisms. The organisms that thrive in the water-mass of an aquatic-system.

Photosynthesis. The process of synthesis of organic compounds using the energy in sunlight, through the chlorophyll molecule, according to the general equation: $CO_2+H_2X \rightarrow (CH_2O)+H_2\ O+2X$. In oxy-photosynthesis of some bacteria and all photosynthetic eukayotea, X= Oxygen; in anoxy-photosynthesis of some other bacteria, X is never oxygen- it is often sulphur, $H_2\ X$ then equalling H_2S.

Phylogeny. The stages of development of any group of organisms, which it has crossed to reach to the present stage of evolution.

Plankton. Pelagic organisms that effectively are suspended in water and can not swim or progress against the flow of water.

Range zone. Represents the total range of a particular useful species/genus or any other element of the fauna in a stratigraphic sequence.

Recurrent fauna. The assemblage, which makes a new appearance after its apparent extinction in lower stratigraphic levels.

Recycled fossils. The fossils of older rocks re-deposited in younger sediments/rocks.

Relict fauna/fossil. The skeletons of organisms released in the Recent sediments due to weathering of fossiliferous rocks.

Seismic Stratigraphy. Study of stratigraphic and depositional-facies as interpreted from the seismic-data.

Sequence Stratigraphy. Depositional sequences, that are packages of strata bounded by unconformities that has gained particular prominence since 1970's.

Stage. A part of a series that was deposited during the geological age.

Sessile. The benthic-organisms that are permanently attached to the bottom of an aquatic system; not capable of locomotion.

Sexual reproduction. A kind of multiplication in which there is exchange of chromosome material during meiosis, and in which gametes combine in the process of fertilisation.

Somatic. The body as distinct from the sex cells.

Stenohaline fauna/species. A group of marine organisms/organism that live/lives only in the near normal salinity i.e. 35 ‰ and are susceptible to change in salinity.

Subzone. Local divisions of a biozone where the stratigraphic units are arranged in the order of superposition. Subzones of one area may not synchronise with those of the other area.

Supergroup. The association of mutually related groups.

Synecology. The collective relationship of organisms to their environment in which they occur naturally.

System. The part of an Erathem, represented by a succession of rocks, which were deposited during a **period**.

Thanatocoenose. The total assemblage of fossilised organisms, which is found in any stratigraphic unit, irrespective of their being **endemic** or **disjunct** to the place of their deposition. Or the total fossil assemblage in a unit of organisms, which existed during a certain phase of geological history and which, thrived in different **biotopes**.

Thanatotope. The place where a **thanatocoenose** occur.

Zone. The smallest unit of the stratigraphic scale (Term subzone has been introduced for specific conditions).

CHAPTER - 2

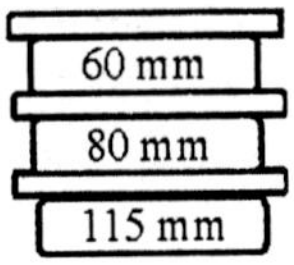

Techniques in Micropalaeontology

Contents

2.1 GENERAL ASPECTS

Micropalaeontology is a technique-based division of Invertebrate Paleonotology. Microfossils being small and fragile require a 'skilled-hand' and 'appropriate method(s)' for collection of samples, their separation from rocks, studies, illustrations, statistical treatments etc. It is generally not possible to separate all the skeletal materials contained by the rock and prevent their breaking during the process of separation from rocks.

The methods of separation of the different microfossil groups not only depend upon their type and composition but also the type, composition and compactness of rocks/ sediments sampled. Foraminifera are no exception to this. They are commonly transported by the currents and are found fossilised in most of the sedimentary rocks, particularly in clay and marl. They are generally scarce in coarse sands but often abundant in silts and fine sands, especially if glauconitic. Many limestones are built of foraminifera but it is difficult to separate them from such rocks due to their dolomitization even at its earlier stage.

2.2 COLLECTION OF SAMPLES

All the known fossiliferous rocks may be the source of microfossils. Normally microfossils are rarely seen in hand-specimens of rock so the only guide of their presence will be the lithology and presence of abundant megafossils in the field indicating that the environment of deposition was favourable to abundant life. They may also be observed in the weathered-surfaces as well as on the bedding-planes. Larger foraminifera may be picked up or sieved out of sediments on talus slopes. For the other microfossils samples of rocks may be collected after removing the weathered part.

Samples may be collected from subsurface-borings at a desirable interval. The majority of oil-wells are drilled by the rotary-drill and the microfossils are carried out by the drilling-mud that should be washed to collect rock-cuttings. The lithology/ composition and degree of palingenesis of sedimentary rocks also help in selection of strata for sampling in an exposed sequence. Some of the time-tested and preferred lithologies are: thin clayey or calcareous bands as skeletal material of most of the groups is calcareous; calcareous shale sections in limestone and dolomites for Foraminifera and Ostracoda; glauconitic sandstones; and carbonaceous shale and black shale for conodonts, pollen, spores and pyritized microfossils. In general, medium-grained sandstones and siltstones of arenaceous sequences yield good results as microfossils are also sorted with sediments of similar grain size.

The samples from exposed fossiliferous rocks may be collected as 'spot-samples' in case of predetermined stratigraphic-levels and as 'channel-samples' (multiple samples in a row) each litho-unit of large stratigraphic-sequence with a constant interval varying from 1 to 3 m depending upon the thickness of the sequence as shown in Fig. 2.1. For 'fresh samples', however, borehole-samples up to the depth of 6000m may be collected.

To avoid contamination, a sample (usually 500 g) is immediately transferred in polyurethane sample-bags, with proper labeling.

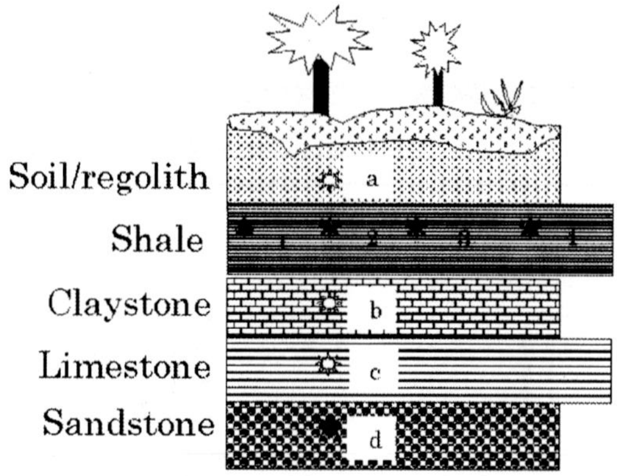

Spot-samples a, b, c, d (one from each litho-unit) and channel-samples (1-4, from a single litho-unit)

Figure 2.1 Spot and channel-sampling in a litho-section

Samples may be collected from the sea as core/grab samples. In grab-sampling generally the top 1 cm layer of the sediments is collected. The samples are stored in plastic/glass tubes with 5-10% formalin to prevent decay of microfauna, which are alive at the time of collection of samples. The location of sample-station with details like nature of sediments, associated macrofauna and flora, water-temperature, pH, etc., are entered in the field-diary immediately. Loose littoral-sediments from beaches may be sampled for the study of modern-skeletal materials prevailing in the surrounding micro-environments.

The wet-samples need to be dried up, in the laboratory before separating micro-fossils.

2.3 SAMPLE PREPARATION AND SEPARATION OF MICROFOSSILS

The rock-types containing different micro-fossils require different methods of treatment, as follows but as a rule less treatment the sample receive the better; microfossils will be less damaged and broken during their recovery.

Their separation from the matrix is performed in a lab having washing-, sieving units, oven, hot plates, fume-cupboard, petri-dishes, beakers, pastel and mortar, filter papers and a physical-balance. A sieve set with 60, 80 and 115 mm mesh sieves with a pan should suffice of the variable-sizes of the different groups of micro-fossils (Fig. 2.2).

Rock samples are first broken into pieces of about half an inch diameter with the help of pastel and mortar. Large rock samples are broken in large and strong containers.

The following methods are employed in the given sequence or in other way based on the nature and composition of rock samples:

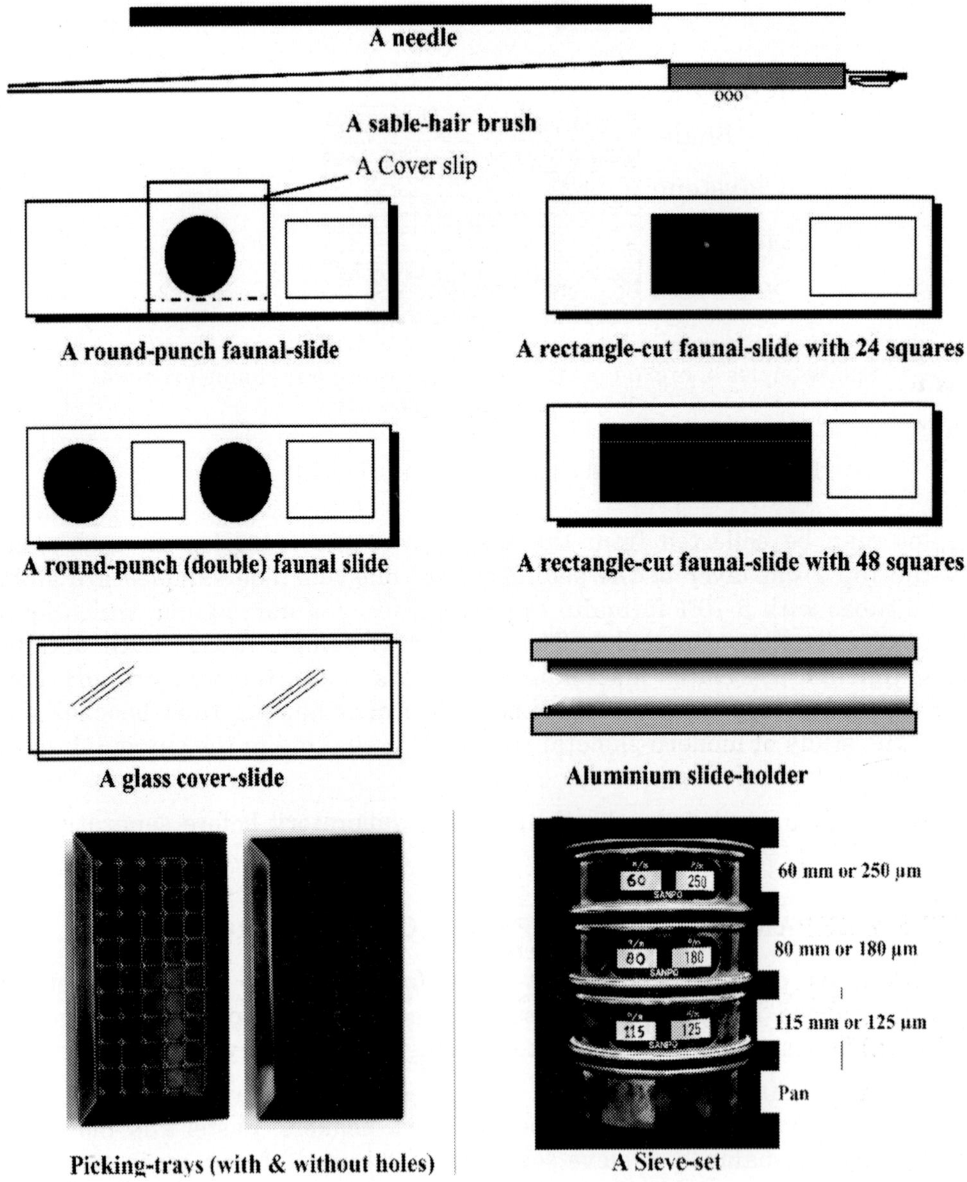

Figure 2.2 Accessories for picking and mounting of microfossils.

Soaking/pulverization in water

A sample (broken in ½ inch pieces) of a definite weight is dipped in cold/hot water for a few hours (for siltstone, mudstone and chalk) or a few days (for limestone, shale and sandstone). This process is slow but yields best results as it prevents breaking of microfossils. Rocks like, limestone or sandstone require treatment in ultrasonic devices.

Pulverization in chemicals

Kerosene or 'turpentine-oil' may be used if soaking in water is not sufficient especially in argillaceous and soft limestone samples. Hydrogen peroxide (H_2O_2) diluted to 0.3% may be used to separate Ostracoda, Foraminifera from mud-, silt, and clay stones particularly if the foraminifera are expected to be fragile. Use of soft-detergents, hard-detergent powders and even caustic soda (Na_2CO_3 or sodium carbonate) with cold or hot water) may be recommended, for half an hour to a few hours, for better results, but with great caution. The ordinary bleaching powder (NaCLO or sodium hypo-chloride) may be used for carbonaceous rock-samples e.g., black-shale, or mudstone where organic walled microfossils are expected. The process applied may be repeated till the sample is fully not pulverised.

Treatment with acids

For resistant rocks, use of dilute HCl may be endorsed if the expected microfossils are chitinous or organically walled (non-calcareous). Acetic acid, oxalic acid and phosphoric acid in desirable concentrations may be approved (with great caution) in case of carbonate rocks consisting of calcareous microfossils. However, chert requires treatment with the fumes of hydrofluoric acid derived from mixing CaF and H_2SO_4 in a crucible in which the sample is suspended. The whole process is performed with utmost care.

Washing and wet-sieving

The pulverised fractions of rocks are then subject to washing with distilled water to remove the parts of clay and silt which. The material is passed through a sieves-set of 60, 80 and 125 mm meshes, corresponding to 250, 180 and 125 μm sizes, respectively with a pan at the bottom (Fig. 2.2). The residue of +60 fraction contains most of the foraminifera and Ostracoda while the +125 fraction is important for pollen and spores, conodonts etc. The sediments and water passing through the 115 mm sieve contains nannoplanktons. The sieved fractions thus obtained are dried in an oven up to 50°C before sorting of microfossils under a binocular stereoscopic microscope.

Alternative methods

If a rock proves difficult to disaggregate the boiling-procedure repeated after the residue has been dried. In many cases mud or shale pallets will be broken down. In rocks like bituminous shale, paraffin or white spirit is used to soak the rock fragments (without warming). The excess spirit then poured off and hot water with washing soda poured on, followed by the normal washing procedure. Some shales will break down if first soaked in dilute hydrogen peroxide (3-15%). The process is speeded up if the sample is hot, but the oxidation-reaction is violent and proper care must also be taken in the task. The Recent sediment samples may be washed with cold water but require staining with 'Rose of Bengal' so that the living organisms particularly the foraminifera at the time of collection of samples can be detected (Murray, 1991), as described below:

1. A 10 ml of sediments is cut and washed clean of mud and preservative under a spray of water over a sieve (mesh size 240 with opening of 63 µm). The volume of the cut may be measured in a graduated vial.
2. The filtrate is transferred to a bowl of Rose of Bengal stain (1g/l) and staining is allowed for 20 minutes with occasional shaking. Alternatively, Sudan Black stain may also be used.
3. The residue is washed under a spray of water to remove the excessive stain from the sediments and shells and then kept of a fine filter paper (Watson No. 42) to allow the shells and sediments to dry.

2.4 STUDY OF MICROFOSSILS

Thin sections

Preparation of thin sections is required especially to study larger foraminifera in hard limestone made in the following way:

1. Cut a parallel-sided slice of about 3 cm long and 2 cm wide by 5 mm thick using a diamond saw.
2. Grind the slice on both sides on 80 grade carborundum on an iron plate.
3. After washing transfer the slice on glass plate with very fine carborundum powder and grind.
4. Fix the slice to a glass-slide with Canada balsam using standard procedure of section preparation.
5. Grind the slice in successive grades of carborundum to obtain a require thinness so that the fauna can be seen in a microscope.
6. Place a drop of Canada balsam on the cover slip and place it on the specimen.

Faunal slides

Separation of microfossils from pulverised rocks needs some accessories, like-finest sable-hair brush (000), faunal-slides made up of card-board with glass-slide and aluminium case (Fig. 2.2).

Sorting of microfossils

Quick method of separation of maximum delicate and air-filled micro-fossils like foraminifera and closed ostracoda carapaces involves treatment of sieved fraction of pulverised samples with Carbon Tetrachloride (CCl_4). The fraction is first heated at 60°C for 10 minutes in an oven and then sprinkled slowly on CCl_4 kept in an open beaker. The air-filled microfossils (due to enhanced buoyancy) of shells keep floating up to 30 seconds. The upper part of liquid is quickly poured on a filter paper kept in a funnel to separate microfossils as shown in Fig. 2.3. The filtrate collected in a beaker is reused. The method is repeated 3-4 times to separate maximum micro-fossils. Considering inflammable nature of CCl_4 the process is performed with utmost care.

The filtrates are collected on the filter paper is allowed to dry and transferred in a small glass-bottle/storage slides, annotated with Sample No., Grain Size (phi)/Sieve Size etc. Use of Tetrabromoethane ($C_2H_2Br_4$) is recommended with utmost care to concentrate conodonts and calcareous or siliceous microfossils.

Picking of microfossils

The various accessories for picking of microfossils and their storage/placement in faunal slides are shown in Fig. 2.2.

The loose and dried material is sprinkled lightly and evenly over pair of chequered picking trays fitted one over the other. The 'upper tray' has holes.

Under the tube of a 'stereoscopic binocular' microscope, all the microfossils are dropped through them into the 'lower tray'. The 'upper tray' is removed and the microfossils are picked up using waxed-needle or a good-quality moist 'sable hair' brush (000) and placed on a tacky surface of a 'faunal slide'/'micropaleontological slides' as shown in the Fig. 2.3.

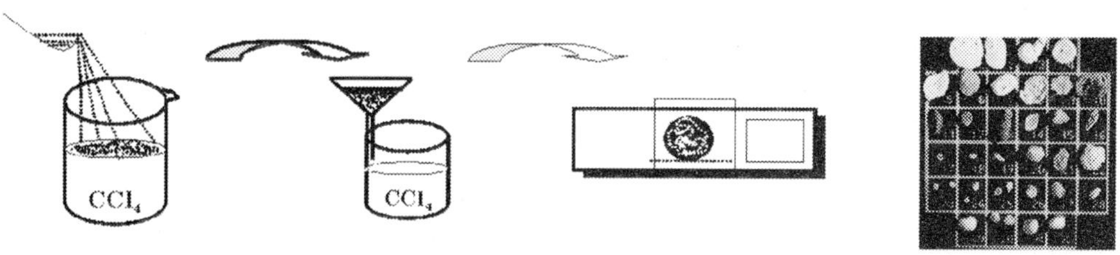

Sprinkling pulverised and dried sample | Filtering | Storage (round-punch) | Faunal - slide (square punch)

Figure 2.3 Use of carbon tetrachloride to separate microfossils and their storage

Different types of 'faunal slides' are available for specific purposes: 'round punch' to store/place skeletons of same species and 'rectangle punch' (chequered with 12, 24, 48 or 100 squares) to store 'type specimens' of one species/all species of a single sample or 'type specimens' of all the species from a single stratigraphic level.

2.5 IDENTIFICATION OF LIVING SHELLS

The protoplasm may remain sufficiently intact to take Rose of Bengal stain up to 60 days after death of organism. As described in the 'Alternative Methods' the ratio of the living verses dead fauna at the time of collection of sediment-samples may be worked out. Such ratios are very useful in understanding the ecological aspects of the microorganisms.

2.6 METHODS OF ILLUSTRATION

Camera Lucida

This simple accessory is used to sketch microfossils. The instrument is based on the principle of superimposition of image of the object, which is viewed (through a prism, fitted inside) and attached on a tube of binocular microscope, as shown in Fig. 2.4.

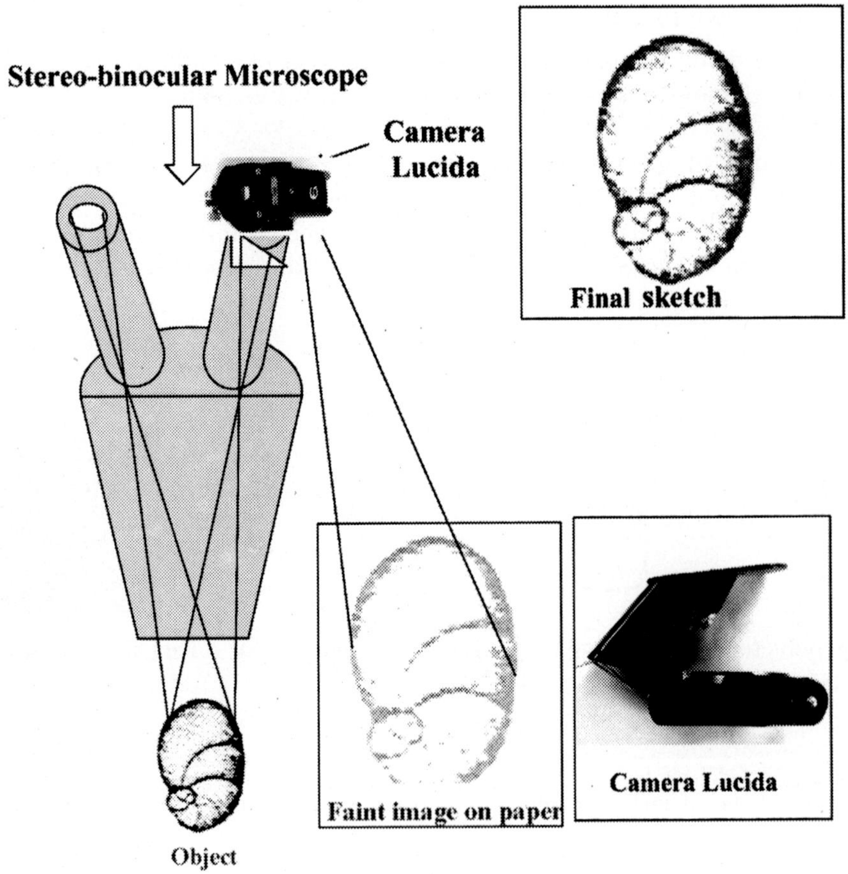

Figure 2.4. Use of Camera Lucida fitted on tube of stereoscopic microscope for sketching of microfossils.

Optical and scanning electron microscope (SEM) micrography

Micrographs may be taken using a 'Close-Circuit' camera attached with a zoom stereomicroscope (Fig. 2.5). The use of electron microscope in the study of micro-organisms and microfossils is carried out through: transmission electron microscope (TEM) using carbon replica, or scanning electron microscope (SEM). The latter is preferred now days as it is easier, faster and, generally gives good results. The selected specimen of microorganism is stuck on a stub of aluminium. The surface of the stub is then coated with a micro-layer (500 Å) of gold/platinum. The stub is then placed inside the electron microscope. A beam of electron scans the object and its visual is displayed on a video screen. The desirable view, angle and magnification (up to 10,000,00 times) may be selected and the object is photographed. Electron microscopy helps in the study of 'wall structure' and 'wall composition' of the micro-organisms significantly, especially foraminifera (Fig. 2.6) and nannoplanktons.

2.7 STUDY OF INTRASPECIFIC VARIATIONS

Intraspecific variations in the species may occur due to host of reasons. However, in some groups like Foraminifera two morphologically different forms exist within a species and this phenomenon is known as dimorphism. In such cases, therefore, study of both the forms belonging to the 'haploid' (microspheric generation) and 'diploid' (megalospheric generation occurs e.g., *Rotorboides granulosum* (Heron-Allen and Earland) from the east coast of India (Kathal and Bhalla, 1998), shown in Fig. 2.7.

2.8 STUDY OF MIGRATORY TREND OF SPECIES

Tracing of migratory trends of species is important to understand paleogeography, paleolatitudinal positions of the sedimentary basins besides their paleoecology and evolutionary trend. Fig 2.8 shows that *Rotorboides granulosum* that originated in the east Pacific, migrated to the Indian Ocean through middle Miocene equatorial currents and remained restricted from 25°N to 25°S latitudinal zone (Kathal and Bhalla, 1996b) could be considered as tropical, epifaunal species. Similar studies on tubulogerinids show that they migrated across the oceans in larval-planktic stages.

2.9 COMPUTATION AND PRESENTATION OF QUALITATIVE AND QUANTITATIVE DATA

After identification of species in an assemblage, the data generated can be utilised in the study of ecological, paleoecological and oceanographical parameters. The taxa are arranged in 'distribution charts' or in 'rank order'. There are various types of 'graph-techniques', which help is the display of the nature and distribution of specimens within the sample or the sampling area e.g., histograms, linear diagrams, pie diagram, maps showing geographical locations of the sketches of species etc.

Boltovskoy and Wright (1976) proposed that the quantitative analysis of benthic and planktic species be done separately because the two groups belong to different habitat and are, influenced by different factors. Recent living and dead forms be also identified separately, as their occurrences are influenced by currents, dissolution, crushing etc.

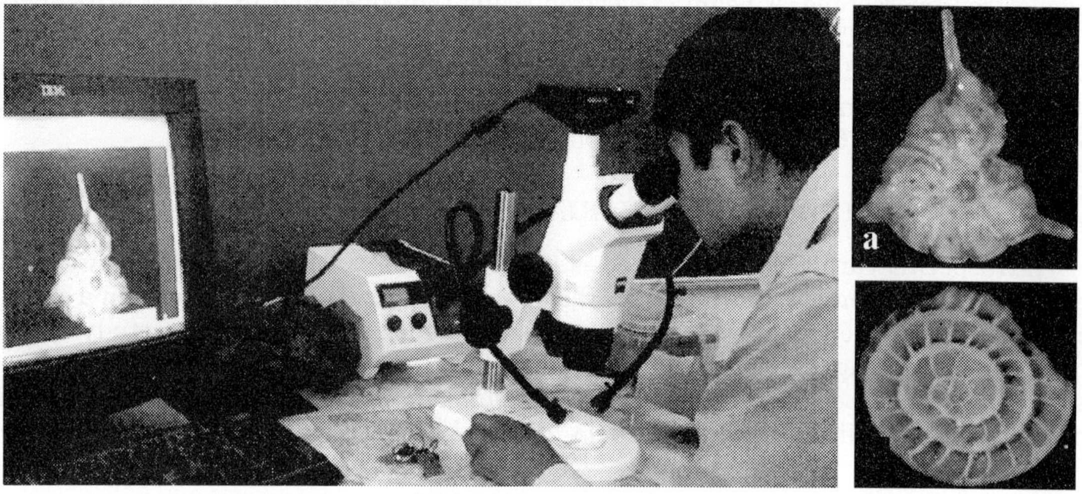

Figure 2.5 Image processing on Close-Circuit Camera- Zoom Stereomicroscope, Micropalaeontology lab, Dr. Harisingh Gour University, Sagar ***a.*** *Asterorotalia trispinosa*, **b.** equatorial-section of *Nummulites*

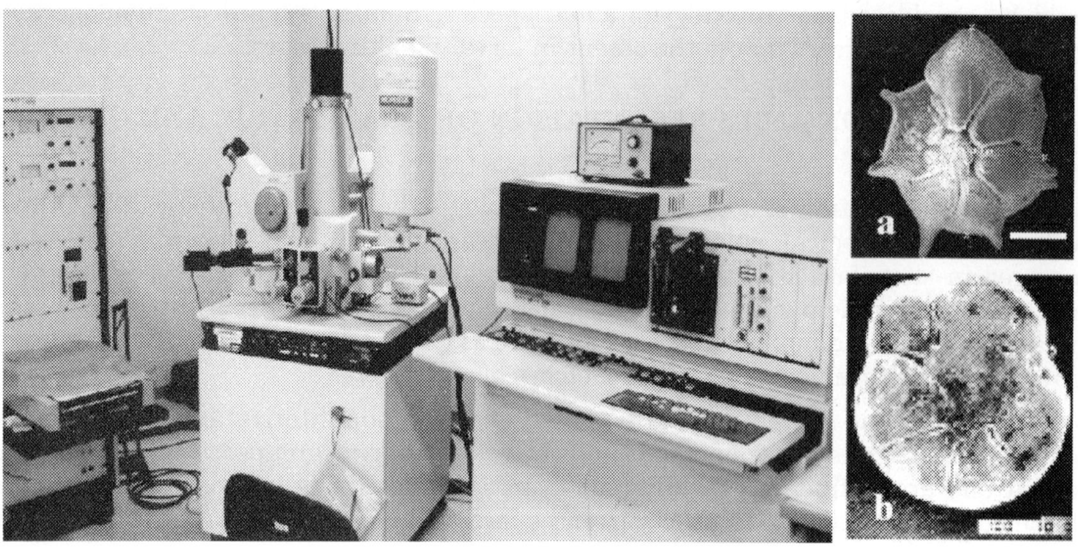

Figure 2.6 Scanning Electron Microscope (SEM) micrographs: a.*Pararotalia*, b. *Rotorboides* (Bar scale 100 μm)

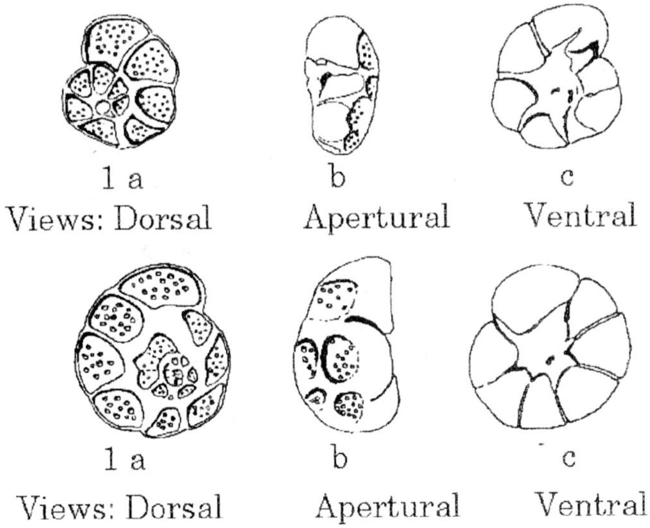

Figure 2.7 Intraspecific variation in *Rotorboides granulosum*:
1 (a, b, c) megaloshperic- and 2 (a, b, c) microspheric- generation (Kathal and Bhalla, 1996b)

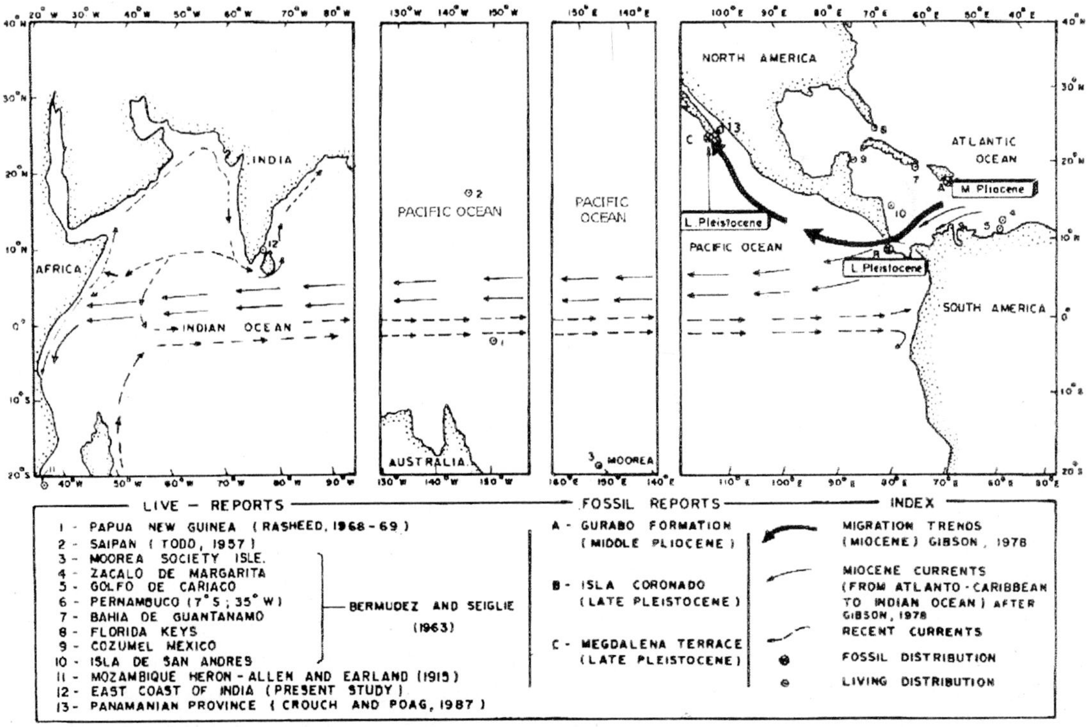

Figure 2.8 Migratory trend of *Rotorboides granulosum*, from the Pacific to the Indian Ocean through middle Miocene Equatorial Currents (Kathal and Bhalla, 1996b)

Statistical methods

Microorganisms lend themselves to a number of statistical analyses, which help in the study of taxonomy, distribution, ecology and population dynamics. Any statistical method requires proper sampling program, which should suite to the requirements of the work and the method. The specimens to be used must always be selected randomly and be abundant. There are several methods of statistical analysis and the selection of one of them is a matter of requirement of the work and suitability of the method. So far the requirement is concerned, the methods may be grouped into two:

The taxonomic determination: In case of morphological variability, the parameters of the variants are first plotted on a graph and then compared to reach to a conclusion of the degree of intraspecific variation possessed by the species. Commonly the digital images are used to handle the large set of data.

Assemblages: Study of an assemblage shows the distribution patterns of fauna and is carried out taking the whole composition of the group based on their taxonomic characters. It is important from several points of views pertaining to the:

i. Autoecology/synecology of the species in the assemblage;

ii. Identification of biocoenoses from thanatocoenoses;

iii. Faunal associations;

iv. Influence of the environment on fauna;

v. Faunal boundaries;

vi. Bio-/zoo-geographical affinities of the region; and

vii. Composition of faunal assemblages of a region/coast (Fig. 2.9 and 2.10).

The distributional studies

While dealing with a large set of data where separate biofacies are expected quantitative delineation is required. Plotting of the percentage of three principal groups of fauna on a triangular model has helped in delineation of 'fresh water', 'estuarine', 'normal marine' and 'lagoonal' fauna (Murray, 1973) as shown in Fig 3.16). Other purposes like the population-diversity and biofacies may be met by an appropriate statistical-technique. Several multivariate analyses are used to define 'species variation' and the 'clumped nature of micro-organism's distribution', have been used successfully.

The popular analyses among the numerical taxonomists are:

(a) Cluster analysis (CA) is a well understood method of statistical treatment to segregate entities (samples, species, measurements) into 'naturally occurring' groups, and quantifies the relationship occurring between groups.

(b) It is also used to judge the degree of similarity prevailing in the fauna. The similarity ranges from '0' (completely dissimilar) to '1' (completely similar) assemblages. Such values of similarities are arranged in hierarchical order until the clusters of similar samples are formed.

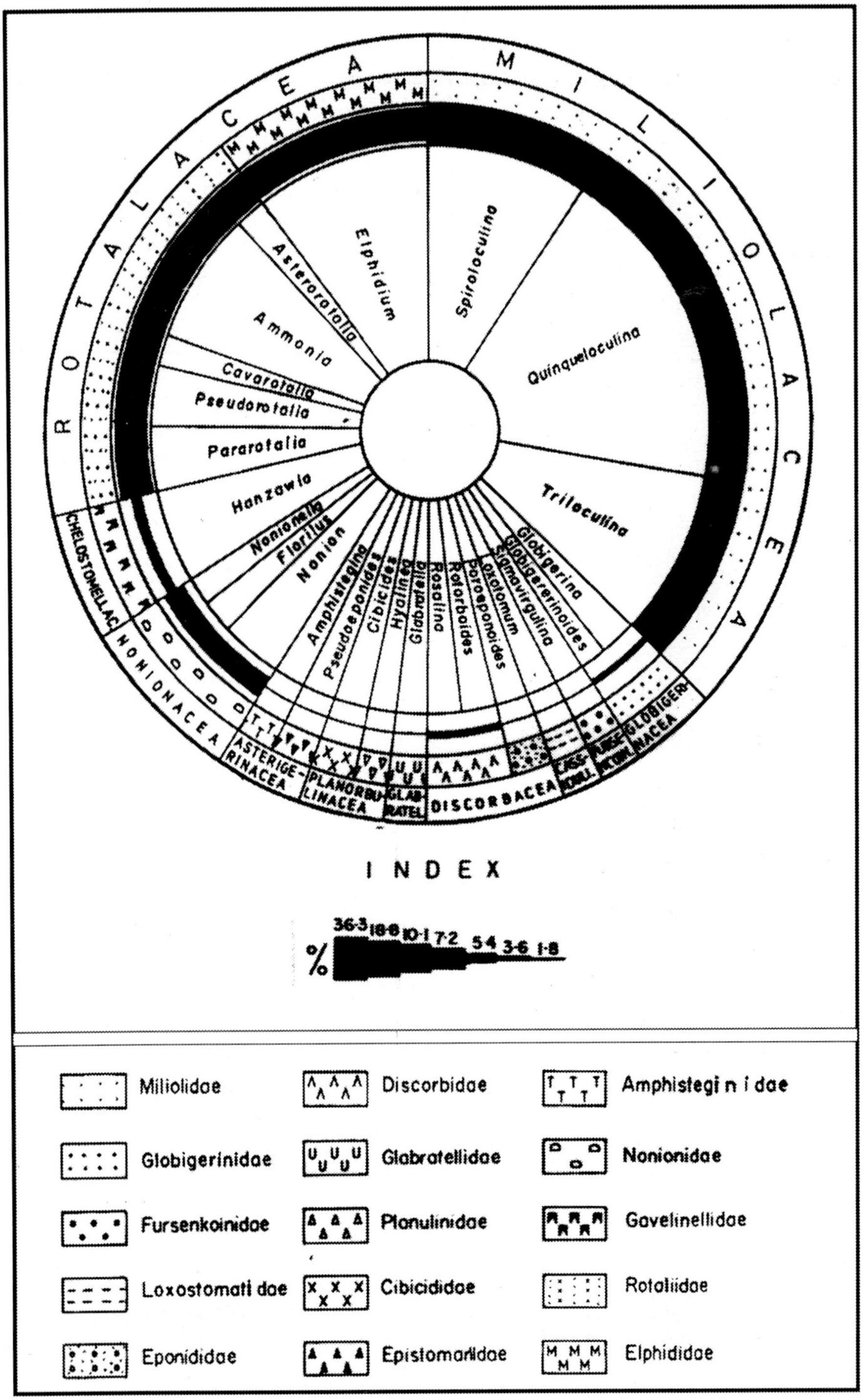

Figure 2.9 Conventional way of illustration of faunal-composition based on Recent foraminiferal assemblage, east coast of India (Kathal, 1989)

Suborder	% Super	Superfamily	Family	Genus	Nos.	Genus %
Textulariina		**Textulariacea**	Textulariidae	***Textularia***	3	
Miliolina		**Miliolacea**	Spiroloculinidae	***Adelosina***	1	
				Spiroloculina	6	
			Hauerinidae	***Quinqueloculina***	11	
				Triloculina	5	
Globigerinina		**Globigerinacea**	Globigerinidae	***Globigerina***	1	
				Globigerinoides	1	
Lagenina		**Nodosariacea**	Lagenidae	***Lagena***	1	
			Ellipsolagenidae	***Fissurina***	1	
				Oolina	1	
ROTALIINA		**Bolivinacea**	Bolivinidae	***Bolivina***	2	
		Loxostomatacea	Loxostomatidae	***Loxostomum***	1	
		Buliminacea	Uvigerinidae	***Uvigerina***	1	
			Reussellidae	***Reussella***	1	
		Fursenkoinacea	Fursenkoinidae	***Sigmavirgulina***	1	
		Discorbacea	Eponididae	***Poroeponides***	1	
			Rosalinidae	***Rosalina***	2	
				Rotorboides	1	
		Glabretellacea	Glabratellidae	***Glabratella***	1	
		Planorbulinacea	Cibicididae	***Cibicides***	1	
				Lolatula	1	
			Planorbulinidae	***Planorbulina***	1	
		Asterigerinacea	Epistomariidae	***Pseudoeponides***	1	
			Amphisteginidae	***Amphistegina***	1	
		Nonionacea	Nonionidae	***Nonion,***	3	
				Nonionellina	1	
				Nonionoides	1	
				Pseudononion	1	
		Chilostomillacea	Gavelinellidae	***Hanzawaia***	4	
		Rotaliacea	Rotaliidae	***Pararotalia***	3	
				Ammonia	4	
				Asterorotalia	1	
				Pseudorotalia	2	
				Rotalidium	1	
			Elphidiidae	***Elphidium***	8	

Superfamily wt. % 73 % 1%

Species wt. % 11 10 6 5 3 2 1

Figure 2.10 Recent foraminiferal assemblage, east coast of India (Singh, 2009), a digital-illustration of faunal-composition

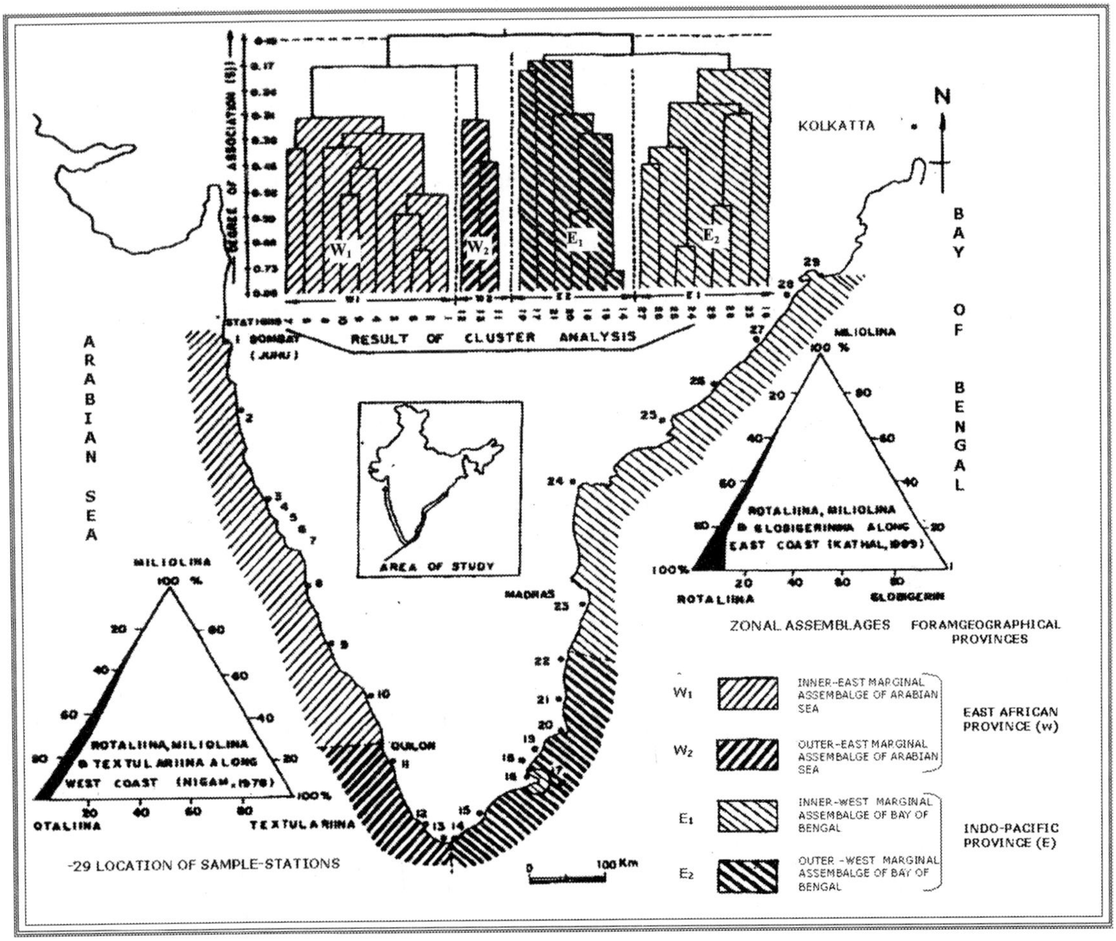

Figure 2.11 Dendrogram (Q-mode cluster analysis), Recent foraminiferal assemblages shows different foramegeographical affinities of the west and east coasts of India (Kathal, Bhalla and Nigam, 2000)

Cluster analysis has been used to delineate bio-facies, species association, and morphometric associations. A drawback of cluster analysis is that it always produces clusters, whether or not any naturally distinct grouping occurs in the data.

The analyses may be formed in two ways: (i) Q-mode for clustering of entities composed of several measurable parameters of one assemblage with other and (ii) R-mode for clustering of variable (measurable parameters) of species with other present in the fauna.

Such studies help understand the 'biotopes' and 'thanatotopes'; solve problems related to the provincial or 'biogeographical distribution' of organisms; and various maritime problems related to sea-food wealth, marine environment and its protection etc.

As an explanatory aid a 'Q-mode cluster' of Recent foraminiferal assemblages, sandy beaches, east and west coasts of India (Fig. 2.11). shows that the 'warm water foramgeographically assemblages' of the west and east coasts of India belong to two different 'foramgeographical provinces': the former to the 'East African Province' while the latter to the 'Indo-Pacific Province', respectively with only 0.17 level of similarity occurring between them (Kathal, Bhalla and Nigam, 2000). Accordingly, as recognized in the areal distribution the dendrograms representing 29 thanatotopes show that all the '13' stations from west coast form a group of cluster 'W' and all the .'16' of from east coast of India form a group of cluster 'E' at the 0.1 level of clustering. Cluster 'W' and the 'E' may be divided into subclusters: 'W1', 'W2', 'E1' and 'E2', respectively.

(c) Association analysis: The 'static chi square' is used as a basis of association between the sample, which are to be clustered in a hierarchical manner.

(d) Factor analysis: An attempt is made to account for the variance between objects with a limited number of 'factors'. The obtained clusters, which respond in the same way, reflect the ecological aspects of the organisms.

CHAPTER - 3

Foraminifera

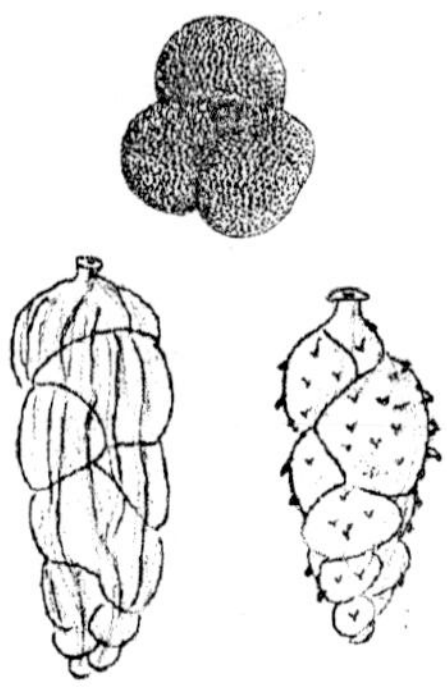

Contents

3.1 INTRODUCTION

Foraminifera (Protozoa) are unicellular organisms with hard shell called 'test'. They are the most diverse group of the modern oceans as living shelled microorganisms and in the rocks of Phanerozoic times as fossils.

They derived their name from 'foramen' the pores connecting the chambers in the tests. They vary in size from less than a millimeter (smaller- or micro-foraminifera) to as large as 100 mm (larger- or macro- or mega-foraminifera) in diameter.

Foraminifera are better known for their spectacular fossil record than for their variety or abundance in the modern marine environments. In 1846, Alicide d'Orbigny, an Austrian scientist counted 68 living foraminifera and estimated that there are some 1000 living species. Zoological Records (1865-1958) revealed that nearly half of the known protozoans are foraminifera. Levine (1962) described about 4,791 living species. Boltovskoy and Wright (1976) estimated that there are about 38,000 fossil and living species known, of which, nearly 10,000 to 12,000 still live in the present-day oceans, sea and lakes while Vickerman (1992) estimated the same is around 10,000 i.e., about one-eighth of all the living protozoans. Loeblich and Tappan (1984) in their latest taxonomic account recognised over 800 genera.

First appeared in Cambrian foraminifera are the enormously successful group of amoeboid protists. They live in the modern oceans right from paralic-environs to both planktic and benthic habitats of the deep oceans. They also occupy epibenthic to deep infaunal microhabitats and utilize a diversity of heterotrophic mechanism (Goldstein, 1999) as shown in Table 3.5 and 3.6.

Majority of living foraminifera are benthic with much greater geological record since Cambrian than planktic species, which are around 50 in number and appeared very late indeed. Long geological history, wide geographic distributions, exceptional diversity in test compositions, forms, and structures and their interpretative taxonomy make them most useful among all marine fossils in stratigraphic correlation, geological age, dating of sediments and interpretations of paleoenvironment.

Being exclusively marine, except a few species, highly abundant in deep-sea sediments/rocks, very sensitive to even minute change in environment foraminifera have wide applications in stratigraphic correlation, higher resolution biostratigraphy, oil exploration, environmental and paleoenvironmental interpretations. Boltovskoy and Wright (1976) valued their importance for stratigraphers, hydrocarbon geologists, biologists, and oceanographers as below:

(a) *Biological importance*: They are numerous and lend themselves to statistical analysis. There is more than one morphological state per species.

Their highly visible external morphology is sculptured into a complex architecture which yields taxonomic and ecological information. Most of the species are relatively easy and smaller to cultivate in a small laboratory.

(b) *Geological importance*: Because of extensive variability, abundance and rapid evolution, they are excellent biostratigraphic indicators.

Being microscopic they are highly susceptible to small environmental changes and are found useful in the studies paleoenvironmental and paleoecological studies. In the last 60 years their study has made significant contributions in the field of oil exploration. Many specimens (up to 20,000/kg) can be obtained from small samples e.g., bore-hole cores and cuttings.

(c) *Oceanographical importance*: Their study has played a widespread role in the field of oceanography, for they exclusively marine nature and occur in all the possible marine, and paralic water bodies irrespective of geographical and bathymetric boundaries in time and space.

They help in investigation of movement of water masses of the present and past; ecology, paleoecology, paleogeography, climate and paleoclimate including paleomonsoonal precipitation in the last 20,000 years in Asia; opening and closing of straits and land barrier; and marine geology.

3.2 SYSTEMATIC POSITION AND IMPORTANCE

Superkingdom............. Eukaryota
Kingdom.............. Protista
Phylum........... Protozoa
Subphylum... Sarcomastigophora
Class..... Sarcodina
Order Foraminiferida

Eukaryota: Unicellular and multicellular organisms with nucleus;

Protista: Unicellular organisms with nucleus;

Protozoa: (*protos,* first; *zoon,* animals) Unicellular, heterotrophic animals with nucleus;

Sarcomastigophora: (*sarcodes*, fleshy, *mastrix*, whip; *phora*, bearing) Organisms of fixed shaped, locomotion by whip-like flagella or finger-like pseudopodia;

Sarcodina: fleshy protozoans of fixed shape; and

Foraminiferida: A cytoplasmic body, enclosed in a test of one/more interconnected chambers, wall proteinaceous/agglutinated/secreted (calcite, aragonite or rarely silica); calcareous walls may be porcellaneous, microgranular or, hyaline; alternation of generations; one or more apertures in addition to the wall perforations through which pseudopodia protrude; benthic or pelagic, may live in marine, estuarine or rarely in fresh water, Cambrian to Recent.

Their cousins in Class Sarcodina are: amoeba (unshelled) belonging to Amoebida) and thecamoebians (sac-shaped tests of chitin-like substance) belonging to Thecamoebida and Radiolaria (siliceous test).

3.3 THE LIVING ANIMAL

As mentioned earlier also they are exclusively marine (except a few allogromids), unicellular animals with secreted or agglutinated test. The cell has a continually changing amoeboid body. Although small in size (commonly less than 1 mm, rarely up 12 cm) they are highly abundant (up to 2600 living specimens/10 sq cm) on the sea floor.

By and large they are omnivorous. However, several 'phytal' species live closely associated with marine algae where the carbon dioxide is utilised by the algae and the released oxygen is consumed by them. The relation helps them attain large size.

Some 'endosybiont species' allow microscopic algae to live below the inner wall of the test that shows a marked 'symbiotic relation' existing between the two. Hallock (1999) found that out of 150 living families of foraminifera, less than 10% harbour algal endosymbionts. Four miliolid, three rotaliid and five globigerinid families host algal symboionts. Besides, several other families sequester chloroplast, which is harvested from algal food, for days to weeks after ingestion. Accordingly, ornamented Peneroplidae hosts symbionts from red algae; unornamented Peneroplidae and Soritidae hosts green algae; Soritinae hosts dinoflagellates and planktic foraminifera host both red-green and green algae as symbiont. Although this is not true symbiosis, yet this phenomenon does enable them to benefit directly from photosynthesis. Three rotaliid families Amphisteginidae, Calcarinidae and Nummulitidae host diatom as endosymbionts.

Cytoplasmic description

The continuously moving cytoplasm is the chief characteristic feature of Order Foraminifera of the Class Sarcodina. The cytoplasm is differentiated into dark-coloured endoplasm surrounded by transparent layer of pseudopodial cytoplasm or ectoplasm.

The cytoplasm takes on different forms in the various chambers of the multi-chambered species e.g., food particles and food residue occupy the last chamber, forming a 'vegetative segment' while all previous chambers contain the nucleus/nuclei and other inclusions. Pseudopodia (*pseudo*, false; *podus* feet), the temporary extrusions from the amoeboid body form a network of ectoplasm in the form of 'filose' (thread-like), reticulose (branching) or anostomosing envelop of 2 to 10 times the diameter of the test (Fig. 3.1). They serve as sensory devices in capture and for ingestion and digestion of food, in the test besides temporary cyst-construction, for anchorage in locomotion. Pseudopoda-net may rise from apertural-region of the test alone in tectinous, porcelaneous, and agglutinated species, or may radiate through many tiny perforations in of hyaline (perforated) tests. The ectoplasm is intimately involved in the secretion of test (calcareous forms). It has a remarkable ability to dissolve the calcareous material on its external surface and secrete test material on its inner layer (Banner, 1971).

The endoplasm, in a single chambered form is more or less homogenous with uniformly distributed inclusions. Prior to the reproduction, particularly the sexual reproduction, the entire endoplasm often under goes a process of 'purification' in which all the waste material and pigments are expelled beyond the test and the endoplasm becomes white and translucent (Boltovskoy and Wright, 1976).

Food and habitat

Basically they are omnivorous yet diatoms form their major food, supplemented by other organic materials, like bacteria, phytoflagellates, fragments of larger algae, spores, algal gametes, radiolarians, small crustaceans, nematode, worms etc. To capture food their pseudopodia extends around the food and incorporate them into the ectoplasm where it is digested. The products of digestion are carried inside the test through aperture. They may live in a variety of micro-environments like, brackish water-lakes, hyposaline or normal marine conditions and even in hypersaline lagoons, but most of them live as benthic organisms from river mouths to all known depths (Fig. 3.1c). Some live free and can crawl along the bottom using their pseudopodia as vagrants benthos; others remain attached to the bottoms using pseudopodia as sessile benthos to seed-weeds, corals, molluscs and also inside the cavities of other tests and polychaete worm tubes; and even as parasites in a very few cases. They occur in all latitudes. Some species live in restricted latitudinal zone e.g., *Rotorboides granulosum* lived within 25° N to 20° S latitudes (Kathal and Bhalla, 1996).

Assumedly, larger Tertiary foraminifera, e.g., *Nummulites, Assilina* etc., could attain large size up to 2-3 inch or more due to symboints of different nature with their tests or symbiosis with marine algae.

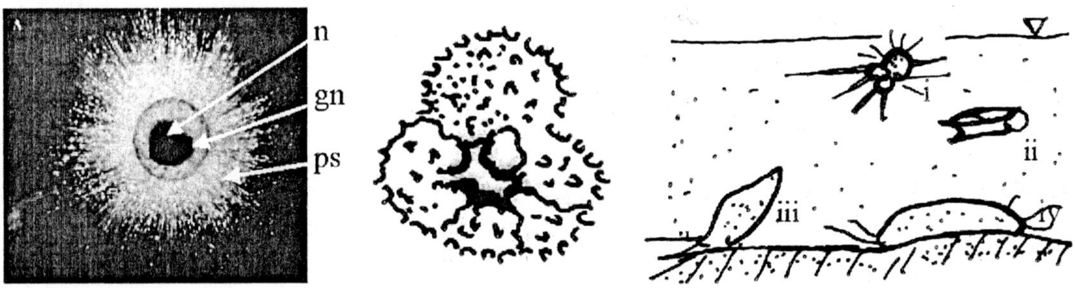

Figure 3.1 a Living Foraminifera (n. nucleus, gn. granular endoplasm surrounded by pseudopodia); **b**. Sketch of planktic species; **c**. Habitat (i. spinose planktic; ii. Keeled-planktic; iii. vagrant benthos; iv. sessile benthos)

A small group of foraminifera is planktic. Some of them have short pelagic stages in various parts of their life history (e.g., species of *Iridia* and *Tubulogerina*). The flagellate gametes (earliest stage) of some of species are pelagic.

Temperatures required both for existence and for reproduction, control their distribution. Any deviation from the normal range may prolong their periods of maturation and may also interrupt the process of alternation in generation. Some species tolerate temperatures from 0° to 34°C while for others the tolerance for optimum temperature slightly differs from that in natural conditions. In general, higher 'species-diversity' (number of species) occurs in warm, shallow waters. The same diversity, however, may be expected in many different environments also.

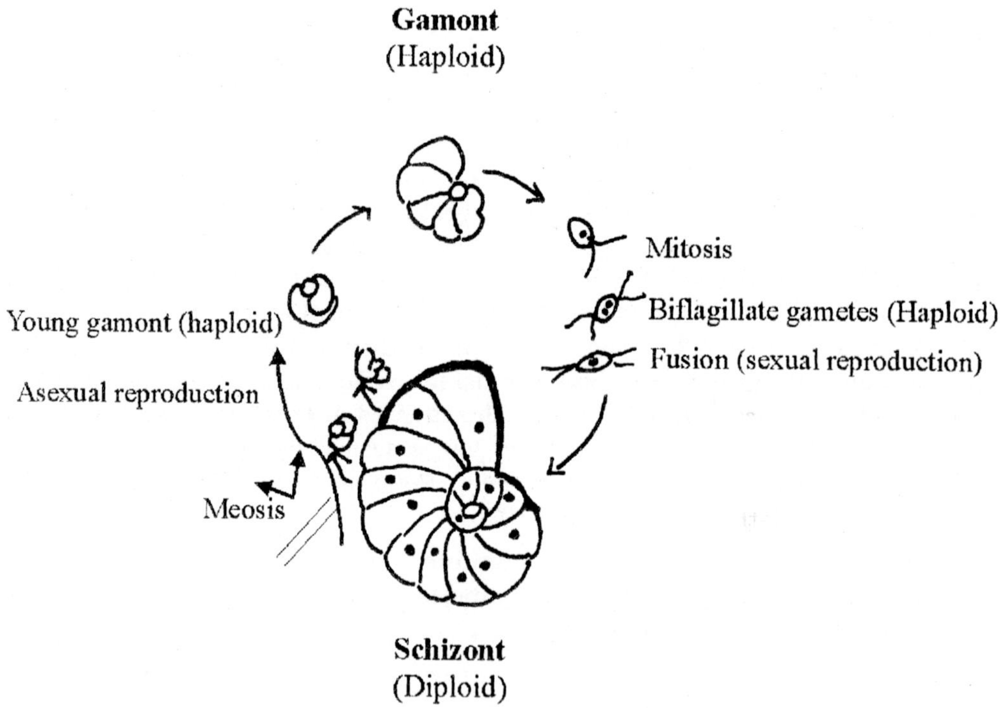

Figure 3.2 Life-history (dimorphism)

Reproduction and life history

Considering the vast number of foraminiferal species little is known about their reproductive habits. Some species show alternation of generations while others reproduce only asexually. Test dimorphism may be present: tests resulting from asexual multiple fissions are megalospheric (large proloculus or 'Form A') while tests resulting from sexual fusion of two gametes are microspheric (small proloculus or 'Form B'), as shown in Fig. 3.2.

In tropical environment the alternation in two generations is completed within a year. In higher latitudes, however, it takes two or more years. In extremely adverse conditions only one generation prevails. During winters, schizont generation undergoes asexual reproduction in a sequence of events: a) withdrawal of cytoplasm in to the test; b) meiosis cell division (or reduction division) in which the number of chromosome are reduced to half, followed by the splitting of the cytoplasm into numerous and tiny daughter cells (gamonts) each with one or more nucleus but half the number of chromosomes; c) process of chamber formation in gamonts begins; and d) the young gamonts are released out into the water for dispersal. On maturity in the summer months the tests undergoes sexual reproduction in a sequence of events: a) the cytoplasm is again withdrawn into the test; b) mitotic cell division (normal division, without further division in the number of chromosome) takes place; c) the cytoplasm splits into tiny daughter cells called 'gametes' (bearing a pair of whip-like flagellum each); d) on their release from the mother cell, two gametes may fuse together to undergo sexual reproduction; and e) their chromosomes are reunited and the young diploid schizonts are released into the water. The entire cytoplasm of the parent individual is involved in both asexual schizogamy and sexual gamogamy, therefore, after the dispersion of juveniles the parent test is often left empty.

Dimorphism

The phenomenon of dimorphism is most pronounced in highly evolved calcareous foraminiferal species then others. The tests of the two generations can be differentiated on the basis of test morphology:

Megaloshpheric test: The gamont stage (haploid, 'X' Chromosome) is recognised on the basis of larger proloculous (first chamber) but smaller test and designated as 'megalospheric' test (Fig. 3.2 and 3.3, 1 a-c and 2 a-c). The 'purinucleate' amongst them are designated as 'A-1' generation and, the 'uninucleate' forms as 'A-2' generation.

Microspheric test: The multinucleate schizont or agamont (diploid, 2X chromosome) stage is marked by a smaller proloculous but larger test size and known as 'microspheric' generation (Fig. 3.2 and 3.3, 3 a-c, 4 a-c and 5 a-c).

Many dimorphs has been mistakenly described as distinct species or even as distinct genera. Modifications of the usual asexual reproduction include shell fragmentation and simple binary fission in some agglutinated types e.g., *Bathysiphon,* or budding and fission in *Halyphysema.* In some orbitoides e.g., *Glabratella* the gametes are triflagellate and are produced by two parent individuals that attach by their umbilical surfaces known as plastogamy to form a joint 'brood chamber' in which the fertilisation takes place. In *Sprillina* and *Patellina,* a few large amoeboid gametes are produced by two or more individuals, which associate to form a joint reproductive cyst. Several other variations in the style of life cyclic have been noted.

3.4 TEST

General description

The foraminiferal cell makes a shell or 'test' of thin, delicate and simple or thick, firm and extremely complex architecture except allogromids that make an organic layer called 'tectin'. It may be homogenous or have two or more similar or differing layers that may have lamellar or labyrinthic structure (blind branching tubes).

The wall may be organic, agglutinated or calcareous. Five possible functions of the test (after Marszalek *et al.,* 1969) are: (i) to protect the cell against predation; (ii) to provide shelter against unfavourable physical and chemical conditions; (iii) to serve as receptacle for excreted material; (iv) to aid in the reproductive processes; and (v) to control the buoyancy of the cell. In addition there are other ways where test can assist growth of the cell like, it protects the cell from certain wave lengths of light in shallow water regimes. In low-oxygen environments, some symbionts are found concentrated below the pores to facilitate as pathways for exchange of gases (Hottinger, 1982, Kathal & Bhalla, 1996, 1998). Species secreting a test by biomineralization is more energy efficient than agglutinations.

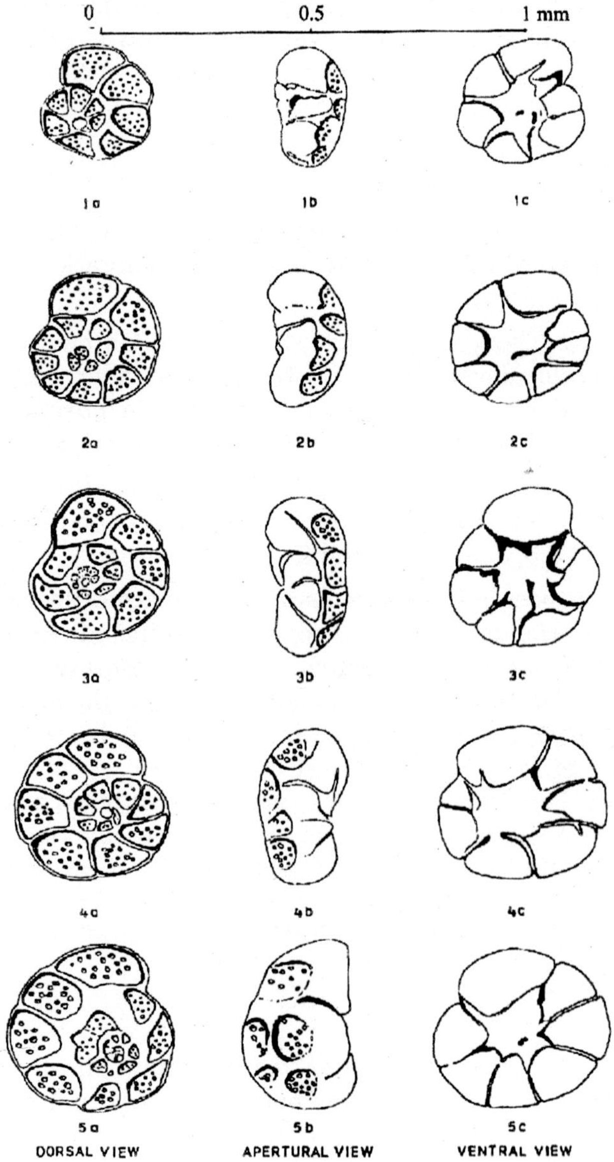

Figure 3.3 Intraspecific variation: 1-2 megaloshperic- and 3-5 microspheric generation, *Rotorboides granulosum*, east coast of India (Kathal and Bhalla, 1996b)

Formation of test also assists the growth of test based on an assumption that there must be an upper limit to the size of a single cell if it is to maintain its integrity (Murray, 1991). Brasier (2008) pointed out that biomineralization involved in the test may be necessary to remove toxic Ca^{2+} from the cell.

Adult tests may range in diameter from 0.02 to 110 mm, but the fundamental unit of a test is a 'chamber'. The wall between two chambers is called 'septum' (pl. septa) whose external expression is known as 'suture'. Growth of individuals may cease construction of test except in a few tubular or branching species where the test may enlarge further in one or more directions. Normally, growth takes place by addition of new chambers, interconnected with the previous ones in continuous or periodic manner. Chambers in different species may be added in a row (uniserial), zigzag (biserial), triserial, coiled along a plane (planispiral) or along a cone (conispiral or trochospiral).

Taking into consideration the majority of all living and fossil foraminifera may be differentiated chiefly and entirely on test sculpture, composition and microstructure (pattern of construction), it has been valued as one of the most important parameters in the most recent attempts to classify foraminifera.

Classification based on the different wall compositions e.g., tectinous, agglutinated, calcareous and siliceous etc., seems to be the most 'natural' one. However, recently sinistral and dextral forms in *Globigerina patchyderma* and *Globorotalia truncatulinoides* have been recognised as two different species based on genomic considerations (Kucera, 2007, *personal communication*). The advent of 'scanning electron microscope' (SEM) revolutionised information on composition and microstructure in foraminifera. Loeblich and Tappan (1964) unraveled their 'phylogenetic relations' on the basis of wall composition and all structure that were otherwise based mainly on morphological aspects like, arrangement of chambers, aperture parameters etc.

Wall composition and microstructures

The various wall compositions are illustrated in Fig. 3.4 and Fig. 3.5 and described below:

Chitinous or pseudochitinous walls

'Chitinous' (a thin, flexible, deformable and transparent) wall, supposedly the most primitive is not represented by fossil record. Awerinzew (1901) checked the validity of chitin in *Peneroplis* and found that the material is rather 'tectinous'. Leoblich and Tappan (1964) included them under 'pseudochitinous walls' along with chitinoid, keratinous and proteinaceous or tectinous varieties.

Only allogromids have tests composed primarily of chitin. In astrorhizids supposedly chitinous layer seems to be concealed under detritus grains, its true nature can not be ascertained fully. Some miliolids (calcareous forms) exhibit chitinous test, if calcium carbonate is below normal in estuarine condition or the pH is low in marsh, bay and lagoon conditions. As a matter of reference all the material similar to chitin has been referred to as 'tectinous' (Adj. tectin) in the forthcoming text.

Agglutinated walls

The electron microscopy reveals that the agglutinated wall of textulariids, considered earlier as primarily made up of arenaceous grains is actually composed of foreign matter attached on an organic lining or held in organic cement, with ferruginous,

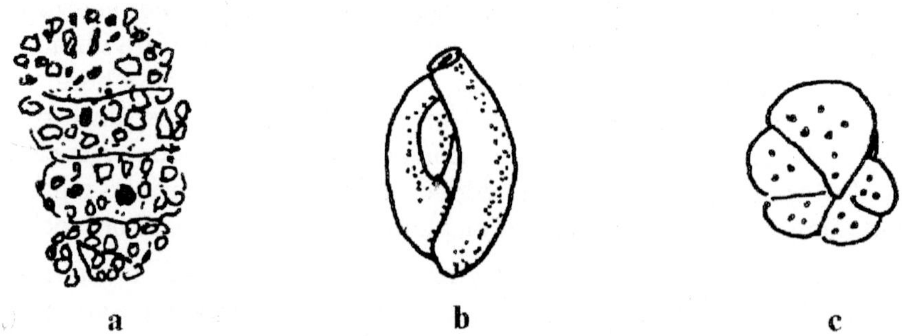

Figure 3.4 Wall composition: **a**. agglutinated (*Rheophax*), **b**. porcelaneous (*Triloculina*), **c**. calcareous perforate (*Cibicides*)

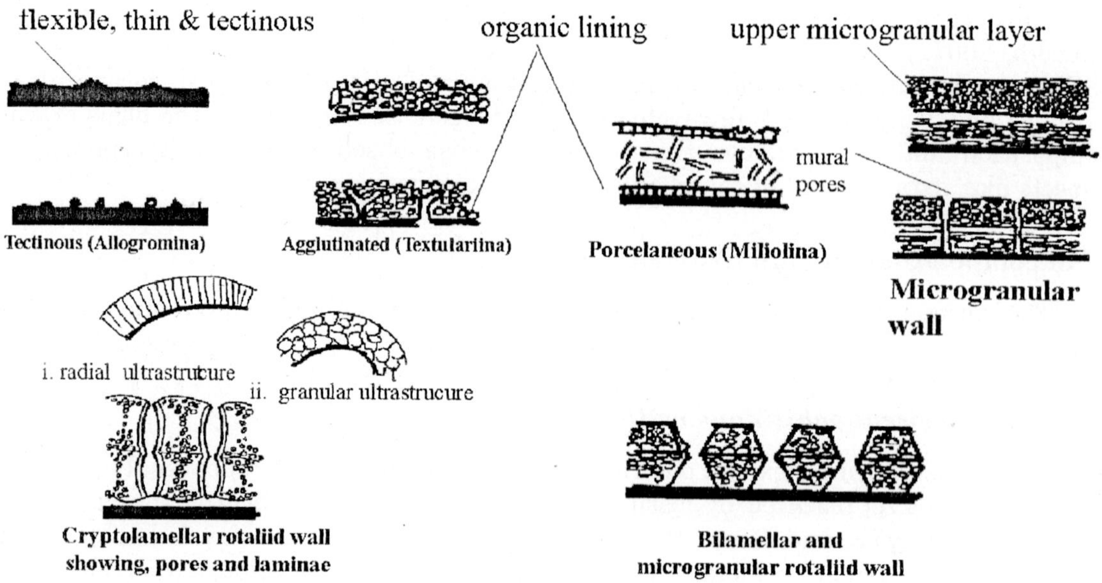

Figure 3.5 Wall structure in some foraminiferal Suborders (modified after Braiser, 1982).

calcareous, or siliceous components. Hofker (1953) regarded the selection of material by the forms as an important taxonomic clue. Studies reveal that species do intend to be selective in the material to be used for test construction and its arrangement, but others utilise whatever material is available in a heterogeneous arrangement e.g., quartz, mica flakes, fragments of clay and carbonates rocks, ilemenite, tourmaline, zircon, volcanic ash, pieces of sponges, pieces of foraminifera, coccoliths etc., and, therefore, the detritus and cement reflect the local environment of deposition.

The grains usually form a highly interlocking mosaic and may have a few alveoli (labyrinthic wall) which are tube shaped structure, closed at the bottom and bifurcating at the top.

All the agglutinated foraminifera are benthic, except *Orbulinoides agglutinatum* (Saidova, 1967).

Calcareous walls

This is the most common of all the secreted walls in foraminifera and occurs in the form of calcite or aragonite (the hexagonal and orthorhombic forms of calcium carbonate). Todd and Blackmon (1956) pointed out that phylogenetic origin of species, rather than the ecology, determines the choice between the two forms of calcite. Also that only robertinids are aragonitic in composition.

The calcareous walls may be one of the following types:

Porcelaneous walls: Williamson (1851) used this term to describe opaque, calcareous, imperforate, milky white walls that shine like 'porcelain-surface' or tooth (Fig. 3.4b). Serova (1961) described it as a 2-15 µm thick wall of three-layers: middle, 2-5 µm in thickness, 'highly magnesian' needles, arranged 'randomly' sandwiched between thin layers of 'orderly arranged' calcite crystals that are perpendicular to wall (3.5). Lack of organisation of internal layer suggests absence of a well-organised organic matrix on which calcite crystals can nucleate. Wood (1963) studied porcelaneous walls in petrological microscope as homogenous and invariably showing low polarisation tints, grey and yellow of first order.

Microgranular/finely granular walls: Palaeozoic fusulinids show a distinct wall, made up of very tiny, equidimensional, subangular and tightly-packed crystals (Fig. 3.5), which appear dark in thin sections and opaque, usually brown and grey in reflected-light. They include some foreign matter also. Secreted secondary deposits in older chambers suggest a secreted origin of the entire test. There are two possible patterns of arrangement of crystals: (i) where granules are regularly aligned, resultant wall is fibrous or pseudofibrous in appearance; (ii) if arranged in two or more layers, the lowest is always fibrous and the upper layer/layers may be microgranular. Some species have perforated or even alveolar walls, having several mural pores or tubular pores.

Hyaline calcareous tests with perforate radial walls: The rotaliids have glassy or hyaline walls (Fig. 3.5). Sorby (1897) found that crystals of calcite in such walls are oriented perpendicular to the C-axes to the surface of the shell. Hansen (1968) noted that since calcite belongs to the hexagonal system, a radial wall is made of crystals whose basal pinacoid (0001) coincides with the plane of growth, i.e. the wall. This orientation gives the wall a glassy transparent reflective appearance. And also, when viewed under the conoscopic conditions, in a petrological microscope they show a 'black-cross' or isogyres and 'rings of colours (interference figure). Krasheninnikov (1956, *vide* Boltovskoy and Wright, 1976) has classified them on the basis of size of the crystals:(i) Finely radial: 0.5-1.0 µm crystals with pores of 1-micron diameter; (ii) Coarsely radial: 2-3µm crystals with pores of 3-4 micron diameter; and (iii) Indistinctly radial: 2-3µm crystal placed vertically and irregularly.

A large percentage of hyaline walls show perforate, radial microstructure (Wood, 1963) and that most of those with perforate granular structure are dominantly members of a few families. Most of the rotaliids e.g., Nodosariacea, Orbitoidacea, Buliminacea, Rotaliacea and globigerinides e.g., Globigerinacea exhibit radial hyaline walls. Intact and small specimens like *Lagena* show a black-cross with concentric rings of colours, similar to the uniaxial negative interference figure of basal section of calcite.

The rotaliid walls are traversed by straight pores or branched-alveoli through which fluids may pass, linking ectoplasm and endoplasm. The pores are provided with perforate organic diaphragms or sieves which act as semi-permeable membrane.

The hyaline radial wall with aragonite crystal arranged in radial manner give an appearance of porcelaneous walls e.g., *Epistomina* and members and robertinids. In some perforate radial walled genera some portions may be imperforate (e.g., keel in *Lenticulina,* septa in *Amphistegina,* secondary deposits in *Cavarotalia, Pararotalia* and ventral side of *Rotorboides* and *Rosalina*).

Hyaline calcareous test with perforate granular walls: It was believed that the hyaline calcareous walls have different optical characteristics when viewed under a petrological microscope. All the foraminiferal tests do not show a characteristically black cross in polarised light (Sorby, 1897). Such walls are constructed of multi-sized calcite crystals, which are randomly oriented optically. This arrangement disperses light rather than reflecting it. However, some of them are transparent because of their thin walls.

Hyaline calcareous test with monocrystalline walls: Test in spirillinds are composed of a single crystal of calcite or more rarely a mosaic of crystals of hexagonal calcite whose 'a-axis' is oriented along the axis of coiling and 'c-axis' parallel to umbilical surface (e.g., *Patellina*) and has been kept in Suborder Sprillinina.

Spicular walls/tests: Tests in carterinids are composed of numerous elongate fusiform spicules each consisting of a single crystal of calcite, with its c-axis parallel to the length of the spicules e.g., *Carterina*.

Radial aranonitic tests: Robertinids have test made up of aranonite, the tetragonal form of calcite arranged in radial manner, i.e. perpendicular to the wall and belong to Suborder Robertinina.

Siliceous walls

Opaline siliceous tests are secreted a very few species belonging to Suborder Silicoloculina. Wood (1949) and Hansen (1972) put forward some doubts on the usage of 'radial and 'granular' terms 'as a major taxonomic tool' to describe microstructure of walls. Further support to this belief came from Towe and Cifelli (1967) when they pointed out that 'the difference in wall structure lie primarily in variation within the stereochemistry of the organic compounds involved in the calcification process'.

Importance in taxonomy

Study of wall-composition and ultrastructure is important for taxonomy, as 'the skeletal features are determined by the nature of secreting protoplasm' (Loeblich and

Tappan, 1964, 1984). The electron microscopy has revealed the true nature of wall composition and structure of both the calcareous and agglutinated tests. These aspects were utilised in distinguishing characters and adjudged as significant for definition of suborders and superfamilies in the suprageneric classification of foraminifera (Loeblich and Tappan, 1964, 1984 and 1988).

Lamellar characters of walls

It is of primary taxonomic importance in the recent classifications (Reiss, 1963) and Loeblich and Tappan, 1964, 1984). Formation of lamella or layer begins after a chamber is formed and commonly remains monolamellar (single-layered) in agglutinated, microgranular and porcelaneous microstructures (Fig. 3.6). In adult 'hyaline calcareous forms' like rotaliids, the walls are multi-lamellar due to a simple reason that each time a new chamber is added, a new layer or lamella is formed in such a way that it covers the entire external surface of the previously formed chambers. So, if a rotaliid test has three chambers it will have three lamellae in the wall of the first, two in the second but one in the third chamber (Fig. 3.6b). Each lamella may in be monolamellar or bilamellar.

Test-form and its morphology

It seems that the oldest of all tests was a 'sac-like' (nearly oval) test with a simple aperture (Fig. 3.7). The protoplasm was able to secret an organic membrane to which fragments of various compositions can adhere; it is believed that first 'unilocular' or single-chambered organism was evolved. Possibly the newly acquired particulate material near the aperture gave the test an elongate form along which further extension of the test could occur. The elongate forms face difficulty in stabilizing in the currents. As a result, more hydrodynamically stable test i.e., a coiled test came into being in the high energy conditions while the elongated forms continued living in calm conditions.

Their adaptation to various environmental conditions and habitat led to a very wide diversity of forms viz. spherical, hemispherical, cylindrical, stellate, flask-shaped, low conical, high conical, fusiform, lenticular, sagitate, palmate, biconcave, discoidal, planoconvex, botroidal, subangular, aborescent, fistulose, flabelliform and tetrahedral. Despite of remarkable variation it is possible to impose a degree of order by recognising that most multilocular test types arise as the result of interaction between three variables during growth; the rate of translation (i.e. the net rate of movement along the growth axis to the net movement away from the growth axis), the rate of chamber expansion, and the chamber shape. The test morphology has been illustrated in Fig. 3.7.

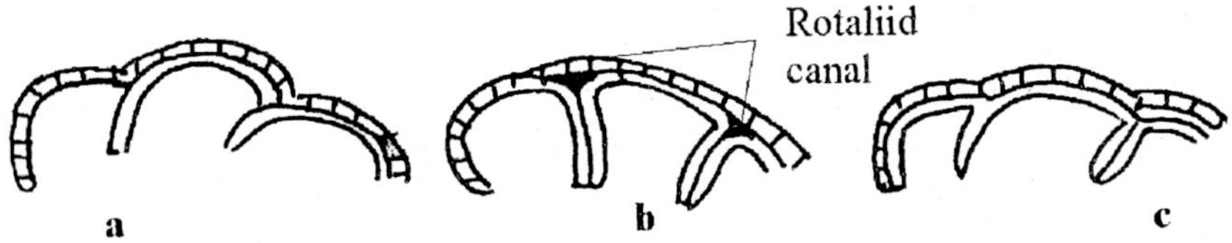

Figure 3.6 Wall construction of lamellar Foraminifera (the cross hatched shows the last formed chamber; a. monolamellar; b. rotaliid; c. bilamellar (modified after: Loeblich & Tappan, 1964).

Chamber-form and chamber-arrangement

Their first formed chamber is known as 'proloculus'. More chambers are subsequently formed on the growth of the organism and the test becomes a multilocular test. Various shapes in chambers are possible: globular, or spherical, ovate, pyriform, tubular, cyclical, hemispherical, clavate, radial, elongate, tubulospinose, angular (conical, rhomboid, truncate) or fistulose.

The shape, morphology and various types of arrangement of chambers are shown in Fig. 3.7, 3.8, 3.9 and 3.10. Different rate of translation may cause formation of planispiral (Fig. 3.9 b, e & h), trochospiral (or conispiral, 3.9 j, k, & l), triserial (Fig. 3.9p & 3.10d), biserial (Fig. 3.9n & 3.10 b) and uniserial (Fig. 3.9 m & 3.10 a) forms. In the planispiral forms the rate of translation remains zero, the chamber or chambers being arranged more or less symmetrically in a plane coil about the growth axis. If the chambers are added in a 'helical-coil' or on an 'inverted-cone' the test is called as trochospiral. Such tests have an 'umbilical' or ventral side (Fig. 3.9l), where umbilicus and only the chambers of the last whorls are visible and 'spiral' or dorsal side (Fig. 3.9j) where proloculus and all the whorls are visible. A successive decrease in the 'spiral angle' in trochospiral test may eventually bring about the reduction in the number of chambers per whorl to three (triserial). The further reduction may obscure or eliminate spiral component, resulting generally in biserial and uniserial growth plans.

The test may be uniserial-rectilinear, uniserial-arcuate (curved), biserial, triserial, biserial-uniserial, triserial-biserial, planispiral (evolute or involute), low or high trochospiral, biconvex, planoconvex, with elevated spiral/umbilical side elevated (spiroconvex/umbilicoconvex), streptospiral or milioline. Milioline tests (Fig. 3.9f, 3.10 q, r, s & t) are tube-shaped, 'half a coil' in length (two chambers per whorl) may be arranged as: (i) Biloculine: chambers added at 180°, embracing earlier whorls completely rendering only last two as visible on both the sides, e.g., *Biloculina* (Fig. 3.10q); (ii) Spiroloculine: non-embracing chambers, added at an angle of 180°, all are visible on both the sides e.g., *Spiroloculina* (Fig. 3.10r); (iii) Triloculine: chambers added at 120°, three are visible externally e.g., *Triloculina* (Fig. 3.10 s); and (iv) Quinqueloculine: chambers added at 144° apart, five are visible externally e.g., *Quinqueloculina* (Fig. 3.10t).

Test openings

Aperture The aperture is the principal opening in the test for extrusion of protoplasm outwards. The position, form and number of apertures may vary from species to species. More apertural openings in the test e.g., 'primary' 'supplementary' and 'accessory' are possible. Similarly, an aperture may have a single or many openings differing in shape and position. So far as the shape of aperture is concerned they may be rounded, subrounded, slit-like, arcuate, radiate, loop-shaped, cribrate, dendritic, cruciform or chevron (Fig. 3.11). The position of the aperture may vary. It may often be situated at the end of chamber (terminal to sub-terminal), with phialine-lip, or may be lateral, interiomarginal (basal), areal, equatorial, peripheral, sutural, umbilical or extraumbilical-umbilical (Fig. 3.12).

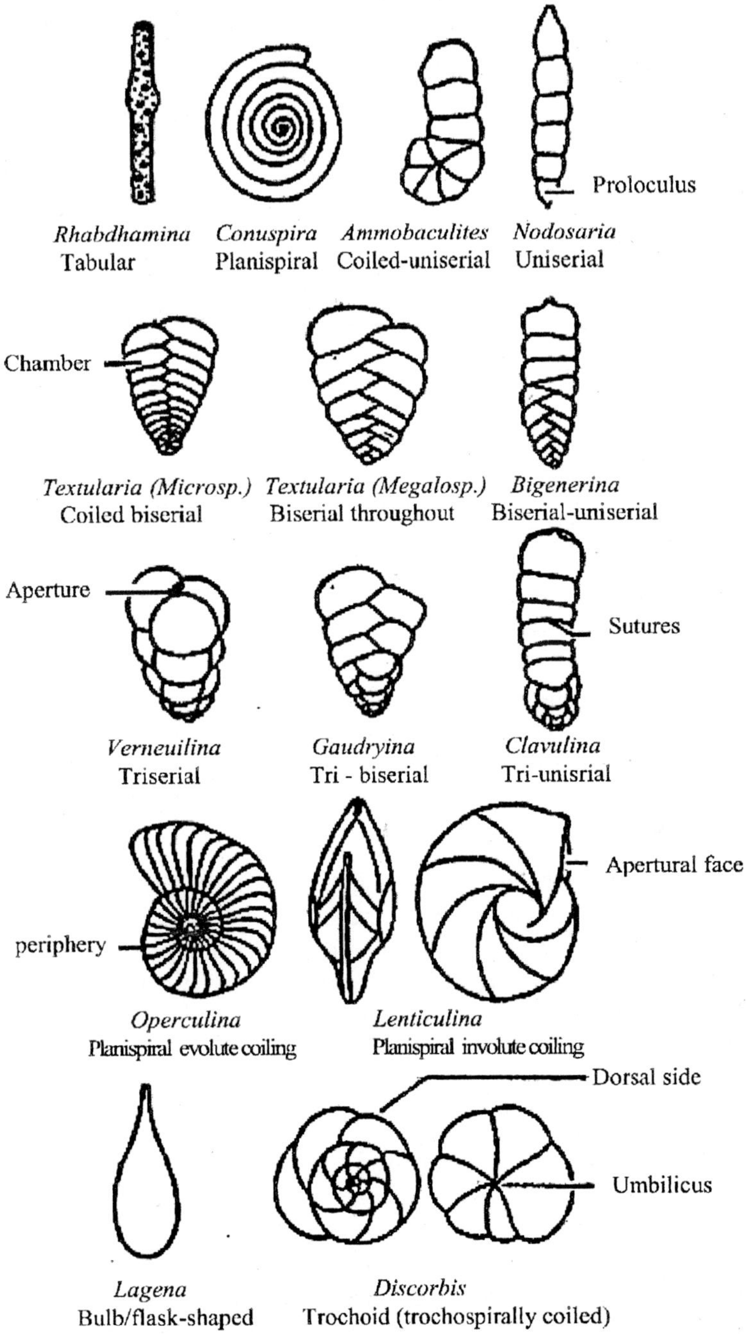

Figure 3.7 Form and structural-features of tests

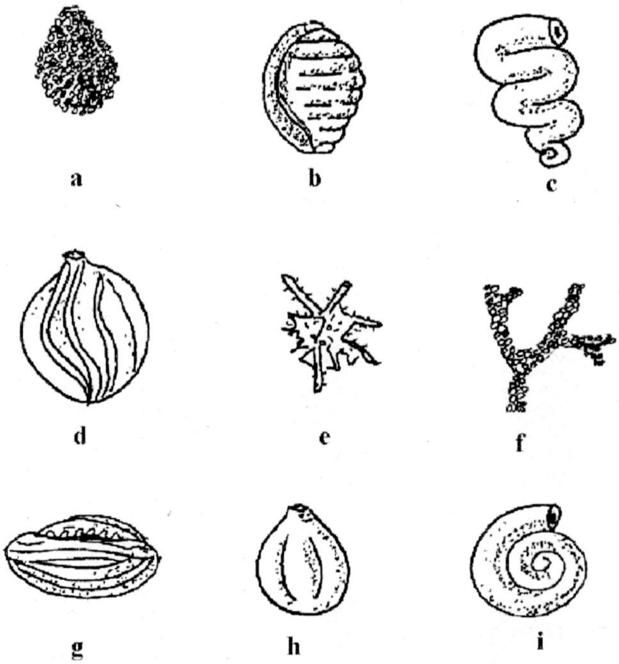

Figure 3.8 Shapes of unilocular tests: **a**. sac-shaped (***Allogromna***), **b**. glomospiral (***Ujbekistania***), **c**. zigzag (***Ammovertella***), **d**. globose or flask-shaped (***Lagena***), **e**. Radiate (***Astrorhiza***), **f**. branched tubular (***Rhabdammina***), **g**. elliptical-spindled (***Fusulina***), **h.** egg-shaped (***Oolina***), ***i***. planispiral evolute (***Ammodiscus***).

Additional openings, formed by the attached portion of the tooth-plate in some species are always found connected with canals in the tooth plate and are called supplementary foramen. By position they may be areal, peripheral, sutural (single or multiple), 'relict' and may be connected with internal partitions or the canal systems. Apertures may also have some external extensions called apertural-lip, apertural-tooth (simple or bifid or complex) or apertural-flap. The aperture is a stable morphological characteristic and has a taxonomic value; however, in some species their arrangement depicts environmental conditions.

Perforations The hyaline and perforated calcareous tests have extremely fine perforations or relatively large or coarse perforations, which may grow, coalesce or join together, or closed with enlargement of the test. They may be randomly spaced or restricted to certain areas of the chambers. Some hyaline forms may remain completely imperforate. More commonly, the apertural lips, tooth plates, surface ribs and carinal bands may be imperforate. Apertures may have 'organic plugs' or 'sieve-pores' in lamellar walls to check infiltration; may penetrate the calcite walls but not the organic lining to allow exchange of gases but no fluids; or organically lined as only a remnant of the organic matrix, strengthening the test by binding the organic membranes of successive lamellae.

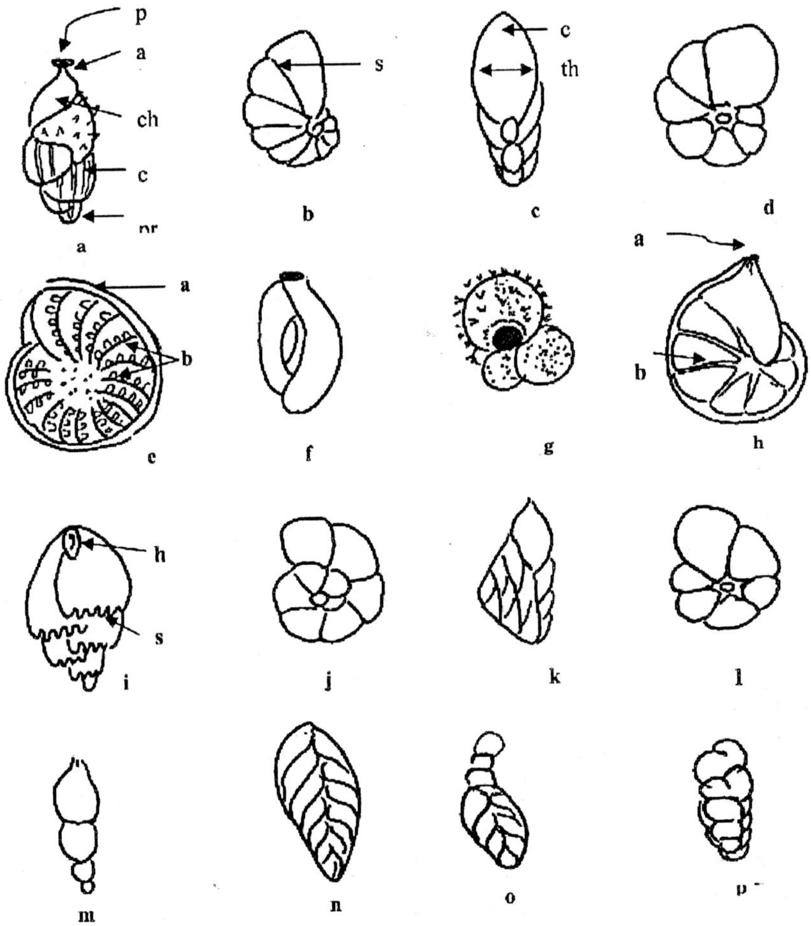

Figure 3.9 Test morphology: **a.** ***Uvigerina*** (p. phialine-lip; a. aperture; ch. chamber; c. costae; pr. proloculus); **b.** ***Nonion*** (lateral view; s. suture); **c.** ***Nonion*** (apertural view; th. thickness; c. apertural-face; a. aperture); **d.** ***Rotalia***, trochospiral, ventral/umbilical view (es. excavated sutures); **e.** ***Elphidium***, a planispiral, lateral view (a. keel; b. retral pores); **f.** ***Quinqueloculina***, 4-chamber view; **g.** ***Globigerina*** (ventral view); **h.** ***Lenticulina***, lateral view (a. terminal, radial aperture; b. keel); **i.** ***Bulimina***, a high-trochospiral (high loop-shaped aperture; s. spines); **j.** ***Ammonia***, dorsal view; **k.** ***Ammonia***, apertural view; **l.** ***Ammonia***, ventral view; **m.** ***Nodosaria***, uniserial-straight; **n.** ***Bolivina***, biserial; **o.** ***Bigenerina***, biserial-uniserial; **p.** ***Verneuilina***, triserial-biserial

Ornamentation

The calcareous tests may exhibit different types of ornamentations in the form of pillars, ribs, ridges, cancellations, keels, flanges, spines, nodes, costae, knobs, retral process and pseudo-reticulations or pseudo-perforations, which are superficial sculptures.

As a matter of fact, ornamentation is a function governed by the environment where the species live. The development of spines in planktic taxa, to increase their buoyancy is formed independent of the primary shell formation. Their size and numbers may vary or be adjusted to cater buoyancy needs of the species in different environments. For e.g., the spinose planktic species live in the upper part of the euphotic zone whereas the non-spinose species commonly live below 50 m depth of water. Cancellate ornamentation makes a film of air around a planktic preventing it from direct contact with water, which otherwise will make the test heavier and eventually drowned.

The ornamentation in benthic or pelagic forms depicts the environment of living. Species of *Uvigerina* may be smooth-surfaced, ornamented with spines distributed randomly/well aligned or costae depending on the depth of water they live in.

Long and equi-radial spines in *Asterorotalia trispinosa* help prevent its burial in soft mud.

The degree or development of ornamentation may indicate chemical environment also. For e.g., below pH 7 the test will bear weak-ornamentation. It may also serve as a 'survival adaptation' in a few benthic species of turbulent waters.

Coiling ratios

Two different coiling directions, dextral (clockwise or right handed) and sinistral (anticlockwise or left handed) may be seen dorsally (spiral side) in some trochospiral tests. The preferences for different coiling-directions have not been understood fully.

A few planktic species e.g., *Globigerina pachyderma* and *Globorotalia truncatulinoides* were believed to show sinistral coiling in cold, temperate and Antarctic waters only.

However, recently, the dextral and sinistral forms have been identified as two different species based on the genomic characters.

The ratio between dextral and sinistral species gives a clue to the palaeoclimate of sedimentation. Changes in coiling ratios have been successfully used for placement of the Pleistocene-Recent boundary.

Nagappa (1957) used the coiling ratio to determine the Lakhi-Ranikot boundary (Palaeocene-Eocene) in Pakistan.

3.5 CLASSIFICATION

> *The science of systematics is divisible into 3 phases:* ***alpha taxonomy****, the first or earlier stage, essentially limited to production of conventional descriptions of species and group of species;* ***beta taxonomy****, the synthetic phase, concerned with proposal and treatment of schemes of natural classification embracing all levels in taxonomic hierarchy; and finally the* ***gamma*** *stage, presumably the ultimate goal of all taxonomy, devoted principally to the problems of evolutionary relationships at intraspecific levels". In general, protozoologist is still in dark ages of stage one, although for practical reasons, he must attempt a bit of beta taxonomy concomitantly*...............Corliss

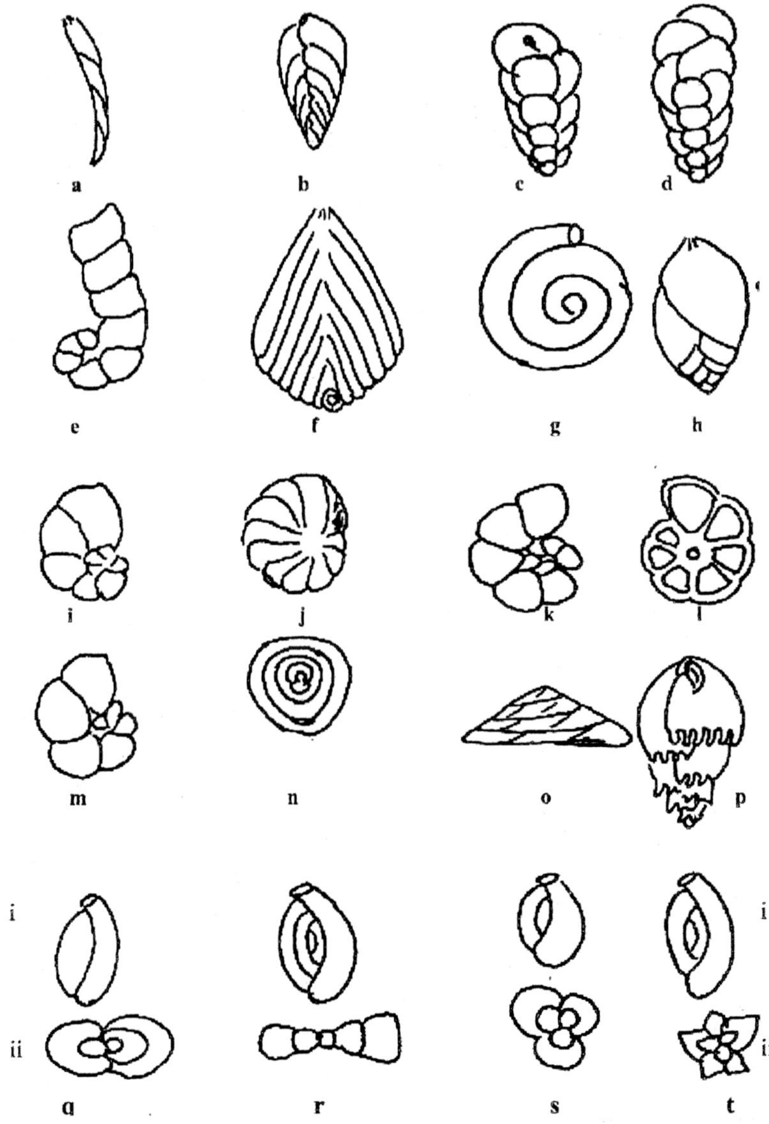

Figure 3.10 Types of arrangement of chambers: **a**. uniserial (***Nodosaria***); **b**. biserial (***Bolivina***); **c**. triserial (***Verneuilina***); **d**. triserial-biserial (***Gaudryina***); **e**. coiled-uniserial (***Ammobaculites***); **f**. flabelliform (***Frondicularia***); **g**. annular (***Conuspira***); **h**. polymorphine (***Polymorphina***); **i**. coiled, planispiral, evolute; **j**. coiled, planispiral involute (***Elphidium***); **k**. trochospiral (***Pararotalia***, dorsal view); **l**. trochospiral, (***Pararotalia***, ventral view); **m**. planispiral, evolute (***Globigerinella***); **n**. annular (***Sorites***); **o**. low trochospiral, side view (***Glabratella***); **p**. high trochospiral (***Bulimina***); **q**. biloculine (***Biloculina***, i. side view; ii. axial section); **r**. spiroloculine (***Spiroloculina***, i. in side view; ii. in axial section); **s**. triloculine (***Triloculina***, i. in side view; ii. in axial section); **t**. quinqueloculine (***Quinqueloculina***, i. in side view; ii. in axial section)

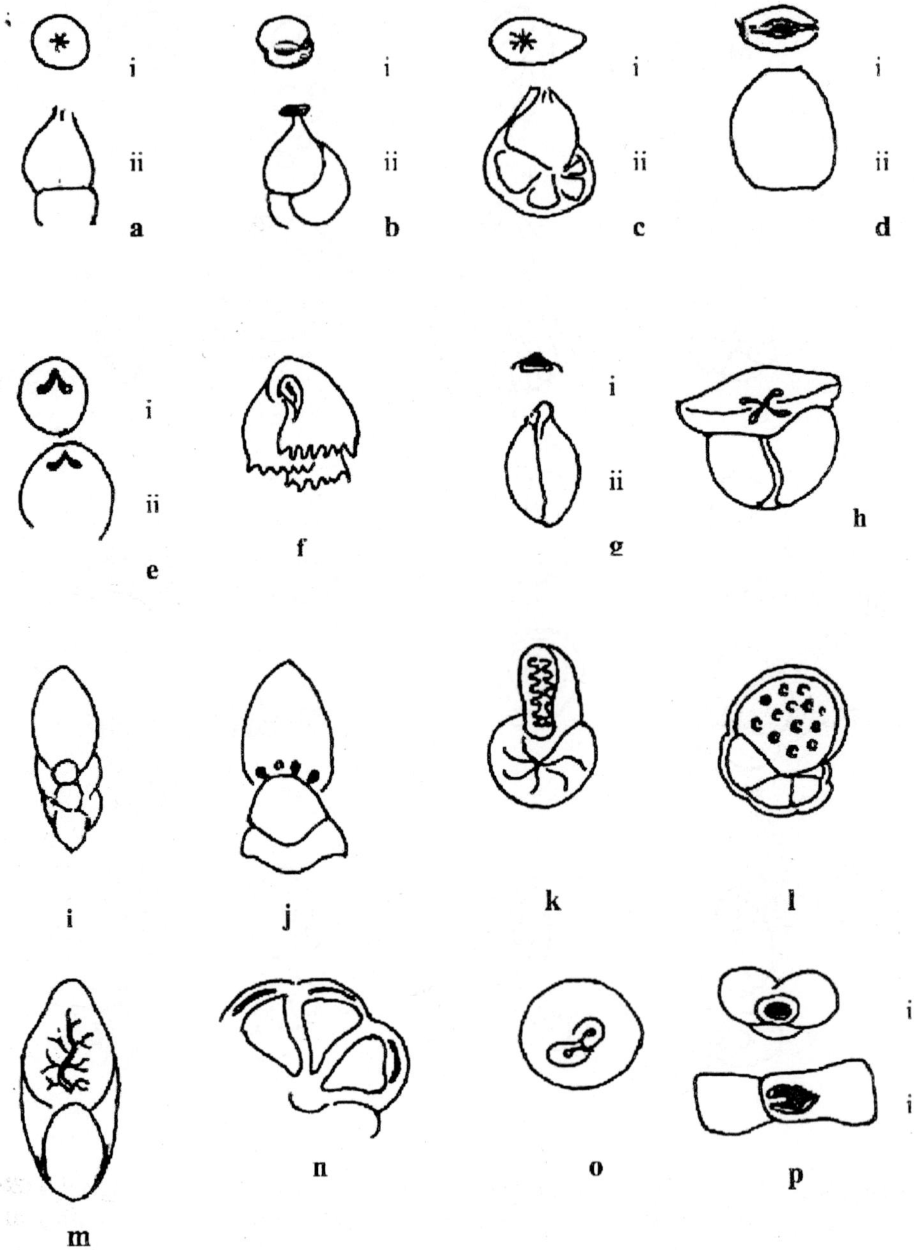

Figure 3.11 Apertures: **a**. radiate (***Nodosaria***, i. apertural view; ii. side view); **b**. phialine (***Uvigerina***, i. apertural view, ii. side view); **c**. radiate (***Lenticulina***, i. distal-end view; ii. lateral view); **d**. slit (***Fissurina***, i. apertural view; ii. lateral view); **e**. chevron (***Ellipsoidina***); **f**. loop (***Bulimina***); **g**. hooded (***Parafissurina***); **h**. cruciform (***Cruciloculina***); **i**. arch (***Nonion***); **j**. rounded multiple (***Elphidium***); **k**. cribrate (***Cribrospiroloculina***); **l**. cribrate (***Poroeponides***); **m**. dendritic (***Dendritina***); **n**. arcuate (***Hoglundina***); **o**. chevron; **p**. rounded, bifid tooth (i. ***Triloculina***, ii. ***Spiroloculina***)

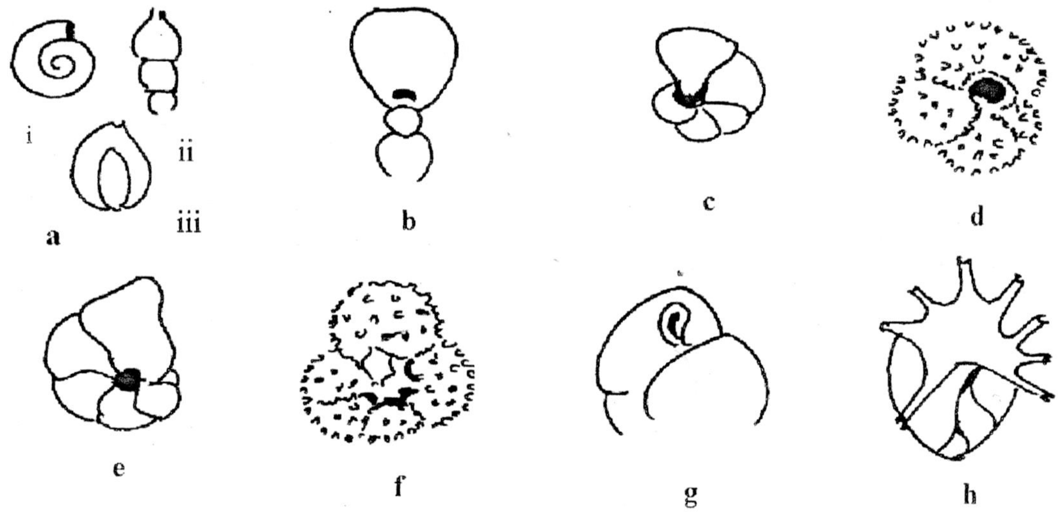

Figure 3.12 Positions of apertures: **a**. terminal (i. circular, ***Cyclolina***; ii. ***Nodosaria***; iii. ***Lagena***); **b**. base of apertural face (***Nonion***); **c**. extra-umbilical (***Globorotalia***); **d**. umbilical (***Globigerina***); **e**. apertural lip (***Globorotalia***); **f**. tegilla (***Trinitella***); **g**.loop-shaped; (***Bulimina***); **h**. polymorphine (***Polymorphina***)

A classification attempts to arrange similar elements of a large system into groups of related characteristics that can be understood and remembered. According to Whittaker a well-known taxonomist, a natural classification should consist of internally coherent taxa, subject to clear definition and delimitation, and based on the consideration of maximum number of characteristics.

However, any attempt to classify the elements of a system of living or dead objects without aim of classification is useless. Moreover, classification of an organism, for the stratigraphical or paleontological purposes, is based primarily on those morphological characters, which are preservable on one hand and reflect the major ontogenic and phlylogenic-relationships regardless of its suitability to biologists.

Cushman (1948) opined that "an ideal classification should be based upon the known phylogeny, of a group as shown by the fossil record, and coupled with the ontogeny of the individual as shown in its complete development, together with what may be learned of the morphology and physiology of the group".

Historical resumé

Many attempts have been made to classify foraminifera by different authors as mentioned below:

Starbo described *Nummulites* from the Egyptian pyramids as a 'freaks of nature', 62 BC-20 AD. Beccarius (1731) illustrated foraminifer as 'micro-molluscs'. Linné also considered them as micro-cephalopods.

Walker (1784), Soldani (1789), Walker and Jacob (1798), Fitchel and Moll (1798, 1803), Lamarck (1801, 1822), Montegu (1803, 1808), and Montfort (1808) contributed research papers on foraminifera before beginning of the nineteenth century.

Five families originally recognised by d'Orbigny (1826); fifty proposed by Cushman (1948), sixty-two by Sigal (1952), seventy-two by Rauzer-Chernousova and Fursenko (1959) significantly had morphological characters as the primary traits of taxonomic divisions. Some of them are discussed here:

De Blainville in early eighteenth-century considered them as cephalopods. Each of the 10 families, assigned to two groups was based largely on the shape of the test.

d'Orbigny (1826), classified them as a foramen-bearing class of Cephalopoda into two suborders: Foraminifera and Siphonifera. He recognised a total of 72 genera. The foraminiferal families were separated based on the shape of the test and arrangement of chambers as uniserial, bi- or triserial, enrolled trochospiral or planispiral and milioline. Order Foraminifera included six Families: Stichostegues, Enallostegues, Helicostegues, Ammonides, Agathistegues, and Enthomostegues.

Dujardin (1835) pointed out that the Foraminifera belong to Phylum Protozoa.

Later, d'Orbigny (1839) elevated the families erected by him in 1825 to the rank of orders based on 'chamber-arrangement'. A few still valid names are: Asterigerinacea, Cassidulinidae, Textulariidae, Polymorphinidae and Miliolidae.

Schultze (1854) classified Rhizopoda into two divisions (?) *viz.* Monothalamia and Polythalamia for the unilocular and multilocular tests, respectively. The former was subdivided into three families (Laginidae, Orbulinidae and Cornispiridae) directly and the latter into three groups: Helicoidea, Rhabdoidea and Soroidea. The Helicoidea had four, whiles the Rhabdoidea and the Soridea had only one family each.

Carpenter, Parker and Jones (1862) divided Rhizopods into two groups: Reticularia, further subdivided based on test structure into 'Imperforata' for pseudochitinous, porcelaneous, and arenaceous families; and 'Perforata' for hyaline perforated families with planktic forms, and the other (unnamed) tests having canal systems.

Reuss (1862) classified fossil Rhizopoda based on (a) wall composition; (b) test construction; and (c) presence/absence of perforations into two groups viz. 'Foraminifera Monomera' and 'Foraminifera Polymera' with seven in the former fourteen in the latter.

Brady (1884) classified foraminifera based mainly on his observations of Recent foraminifera during his famous 'Challenger Expedition'. He discarded the suborders 'Imperforata' and 'Perforata'. Of the ten families Astrorhizidae and Globigerinidae, proposed on the basis of morphological characters and wall structure, are still valid.

Cushman (1925 and 1927), published a 'natural' classification with modifications in Brady's (1884) classification based primarily on 'ontogeny' besides the morphology, wall structure and phylogeny of species. He replaced Gromidae by Allogromidae, and also elevated some of subfamilies to the rank of families. His revised classification (1927) included 15 families of arenaceous forms and 6 of porcelaneous forms. The calcareous and agglutinated subfamilies of the Textulariidae were separated, with recognition of Heterohelicidae, Buliminidae, Cassidulinidae. The Trochospiral hyaline forms were divided into many families and subfamilies. The 10 old families were increased to 45, which included 413 genera. In his book entitled 'Foraminifera their classification and economic uses' (1948), however, he included 50 families and 769 genera.

Galloway (1933) while recognising 35 families realized need of a suprageneric classification. Many features were similar to the Rhumber's classification and recognised Spirillinidae, Endothyridae and Nodosinellidae as separate families.

Chapman and Parr (1936) grouped 36 Families into 3 Superfamilies having 3 different wall compositions one each for the (a) pseudochitinous (for Allogromidae), (b) the perforate hyaline calcite walls; and (c) the arenaceous and porcelaneous walls. The families were similar to the classification proposed by Cushman (1927).

Glaessner (1945) recognised 37 Families, and grouped them into 7 Superfamilies. Two arenaceous superfamilies were erected, one each for unilocular and multilocular tests. Another superfamily included Endothyridae and Fusulinidae. A separate Superfamily included porcelaneous forms. The hyaline families were divided into three superfamilies, first for the Lagenidae and Polymorphinidae, second for the most biserial and elongate genera, and third for all coiled and discoidal forms. Most of the families were similar to the classification proposed by Cushman (1927) except Spirillinidae and Discorbidae, recognised as separate families and Hantkeninidae, Anomalinidae, Rupertiidae and Homotremidae were included in other families.

Later, Cushman (1948) added 5 more families viz., Neoschwagerinidae, Camerinidae, Pegidiidae, Victoriellidae, Discocyclinidae and Miogypsinidae to the 45 of the previous classification of 1927.

Rauzer-Chernousova and Fersenko (1959) directed the publication of a detailed classification of Foraminifera, based on the work done by Soviets. It included 13 orders, 14 superfamilies and 72 families. They added numerous additional families. Ten families of Brady's classification were elevated to the rank of orders.

Classification in the Treatise

The salient features of a classification initially proposed by Leoblich and Tappan (1964) that is based on the wall composition, microstructure and gross-morphology of the tests are:

(a) Basic chemical composition, and radial or granular structure of wall are utilised for distinguishing and defining; suborders and superfamilies;

(b) The cytoplasmic characters, life cycle show good agreement with the classification, although function and origin of many shell characters e.g., canal system, pores, sepal doubling, and apertural tooth plates are yet undetermined; and

(c) Within suborder, the unilocular or multilocular nature of tests is considered important in ranking superfamilies of lower groups (the agglutinated and microgranular forms). This classification included 5 suborders 19 superfamilies 102 families and 1192 genera.

The summarised form is presented in Table 3.1.

Table 3.1 Classification of Order Foraminifera (Loeblich and Tappan, 1964)

Suborder	Superfamily	Nature of test wall	Habitat/Geol. Distribution
Allogromina (organic wall)	**Lagynacea** 2 families, 47 genera	Membranous to tectinous	Marine, brackish, and fresh water; benthic, few planktic; Cam.-Rec.
Textulariina (agglutinated wall)	**Ammodiscacea** 4 families, 84 genera	Simple to labyrinthic	Marine, brackish; benthic; Cam.-Rec.
	Lituolacea 10 families, 209 genera	---do---	Marine, brackish; Miss.-Rec.
Fusulinina (nonlamellar, micro-granular calcite wall)	**Prathuramminacea** 3 families, 23 genera	Simple, undifferentiated	Marine, brackish; benthic; Ord.-Rec.
	Endothyracea 11 families, 72 genera	Granular/ fibrous, 2 or more differentiated layers	Marine, benthic; Sil.-Rec.
	Fusulinacea 4 families, 90 genera	Three or more layers of distinct structure	Marine, benthic; U.Miss. - U.Per.
Miliolina (Porcelaneous, imperforate calcite wall)	**Miliolacea** 7 families, 149 genera	Imperforate wall of tiny disordered calcite rhombs	Marine or brackish Pen.-Rec.
	Nodosariacea 3 families, 86 genera	Radial cal. monolamellar septa	Marine, benthic; Per. Rec.
	Buliminacea 7 families, 68 genera	Radial cal. monolamellar septa	Marine, benthic; Per. Rec.
	Discorbacea 7 families, 68 genera	Radial, perforate, cal. septa, monolamellar	Marine, benthic; Tri.(?) L. Cre.-Rec
	Spirillinacea 2 families, 11 genera	Radial cal., optically a single crystal	Marine, benthic ; Tri.(?), Jur.-Rec.
	Rotaliacea 5 families, 60 genera	Radial cal., sec. bilamellar (rotaliid) canal system	Marine, brackish; U. Cre.-Rec.
	Globigerinacea 8 families, 68 genera	Radial cal., bilamellar septa	Marine; planktic; Jura.- Rec.
	Orbitoidacea 13 families, 93 genera	Radial cal., bilamellar septa	Marine, benthic; Cre. -Rec.
	Nonionacea 3 families, 26 genera	Granular cal., monolamellar septa	Marine, benthic; Tr.-Rec.
	Cassidulinacea 7 families, 34 genera	Granular cal., monolamellar septa	Marine, benthic; Jur.-Rec.
	Anomalinacea 2 families, 32 genera	Granular cal., bilamellar septa	Marine, benthic; U. Tri. (?), Cre-Rec.
	Carterinacea 1 family, 1 genus	Monocry. Cal., spicule in granular groundmass	Marine, benthic; Rec.
	Robertinacea 3 families	Radial aragonite, bilamellar septa	Marine, benthic Tri.(?), Jur. Rec.

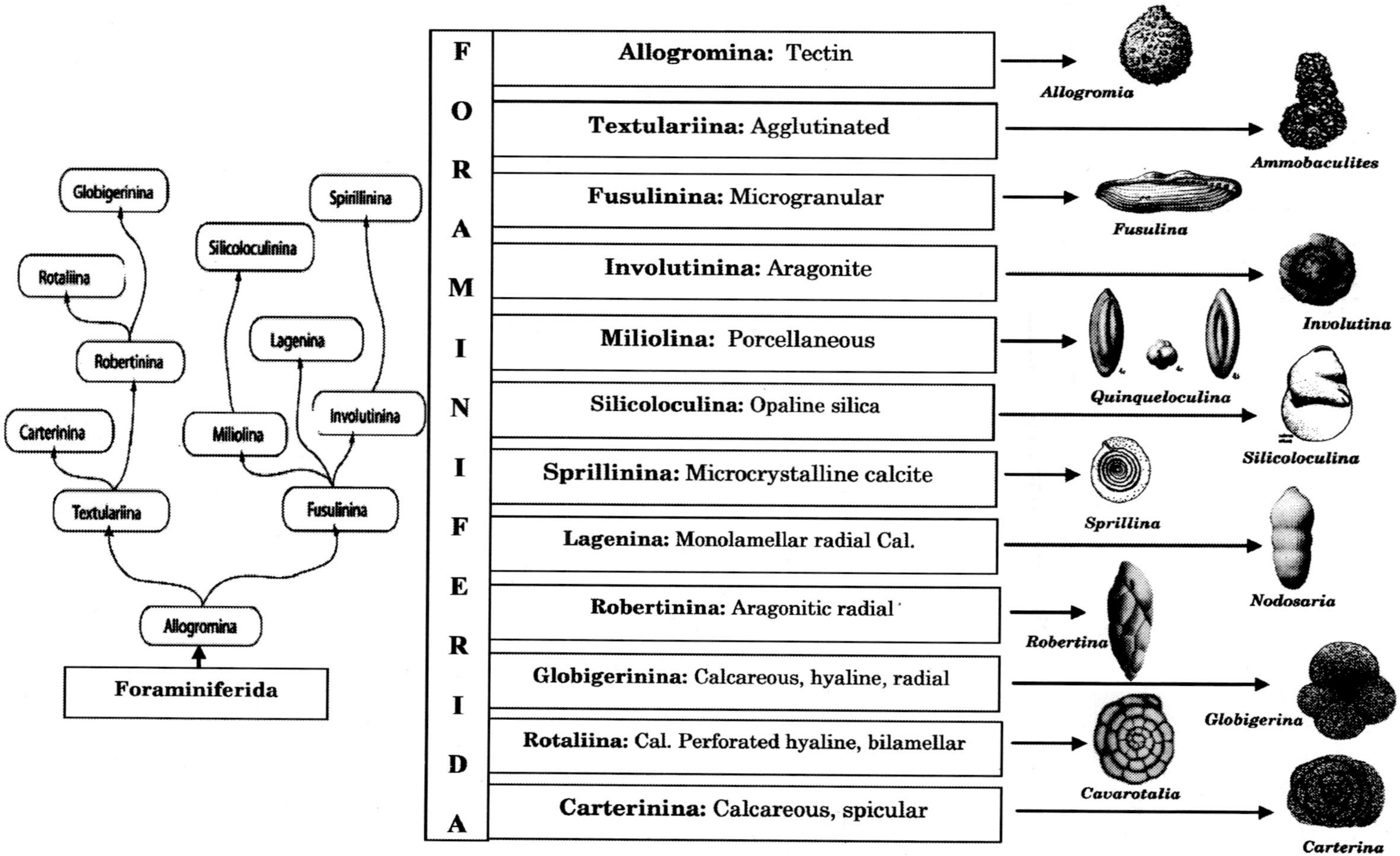

Figure 3.13 Suborders corresponding to the wall compositions lineages and representative genus (based on Leoblich and Tappan, 1988)

As mentioned earlier the use of SEM in understanding cell, test, internal structures, wall and its structure of the species obtained in the DSDP and ODP has added vast data from different facies (carbonate facies of varied ages, deep sea fauna, etc.), different geological ages (particularly the late Palaeozoic and Triassic), and different geographic regions (particularly eastern Asia).

As a result a new classification that can serve biostratigraphic requirements was required.

Some 1150 additional genera were created. To overcome this, the Suprageneric Classification of Foraminiferida (Loeblich and Tappan, 1984) was proposed.

The earlier classifications were based chiefly on morphology in which greater emphasis was laid on test wall composition and structure, as well as apertural feature, and chamber subdivisions.

In the present classification a care was taken by them to not get influenced with the 'apparent similarities in morphological forms which are not related at phylogenic level'.

The total 591 suprageneric taxa were used for the order Foraminiferida included 12 suborders viz. Allogromina, Textulariina, Fusulinina, Involutinina*, Miliolina, Silicoloculina*, Spirillinina*, Lagenina*, Robertinina*, Globigerinina*, Rotaliina and Carterinina* (The asterisked suborders are the new additions to their previous classification of 1964), 63 superfamilies, 253 families and 263 subfamilies.

The salient features are presented in Table 3.2.

Suprageneric classification of Foraminiferida (Loeblich and Tappan, 1984)

Phylum Sarcomastigophora
Subphylum Sarcodina
Superclass Rhizopoda
Order Foraminiferida

The organisms with a cytoplasmic body, enclosed in a test or shell of one or more interconnected chambers; having a wall which may be homogenous, or similar or of layers/laminae which appear differently, or may be imperforate or finely or coarsely perforate; wall composition basically proteinaceous but may have agglutinated particles or may deposit calcite, aragonite or rarely silica, on the organic base, calcareous walls porcelaneous, microgranular or, hyaline and optically or ultrastructurally radiate or granular; sexual and asexual generations alternate, anal or stolon systems of varied complexity may be present; commonly test has one or more apertures in addition to the wall perforations, through which pseudopodia protrude. Sexual and asexual generations alternate habitat benthic or pelagic, marine estuarine or rarely fresh water. Cambrian - Recent.

Table 3.2 Suborder: ALLOGROMINA

Test unilocular, may be multilocular; wall proteinaceous (few agglutinated particles) U. Cam.- Rec.

Superfamily

Nil

Allogromia *Rhabdammina*

Family Characters and Distribution

Families : 5

Lagynididae: Proteinaceous, agglutinations rare; colonial; Rec.

Maylisoriidae: Proteinaceous agglutinations rare; Cam.-Silu.

Allogromiidae: Test globular, ovate, elongate; wall proteinaceous; aperture; U. Ord.-Silu., e.g., *Allogromina, Rhabdammina*

Subfamilies: 3 **Allogrominae**, **Shepheardellinae**, **Argillotubinae**

Hospitellidae: Test multilocular; wall proteinaceous; imperforate; Eoc.-Rec.

Phthanotrochidae: Test uni- or multilocular, forms various; proteinaceous; Rec.

Suborder: Textulariina

Test agglutinated, particles in organic-mineralized groundmass; non-lamellar, Cam- Rec

Superfamily

Astrorhizacea: Test may be irregular, rounded tubular or branching; non- septate; wall agglutinated; Cam.-Rec.

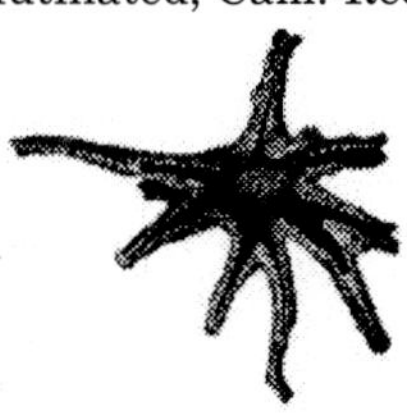

Astrorhiza

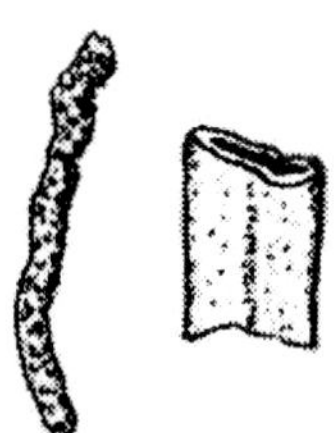

Bathysiphon

Family Characters and Distribution

Families : 12

Astrorhizidae: Test generally large; branched; aperture at the ends of tubes; M. Ord.-Rec. e.g., *Astrorhiza, Bathysiphon*

Subfamilies 2: **Astrorhizinae, Vanhoeffenellinae,**

Bathysiphonidae: Straight, unbranched, non-septate; wall thick; L. Cam.-Rec. e.g., *Bathysiphon*

Rhizzaminidae: Simple or branching, tubular, may be inflated; aperture terminal, on ends; Cret.-Rec., e.g., *Rhizammina* Subfamilies: 3

Dryorhizopsidae: Series of irregular branches; U. Carb.-Rec.

Silicotubilidae: Tubular; pseudo-septate; U. Cret.

Hippocrepinellidae: Elongate, tubular to fusiform, open on both sides, non-septate; U. Carb., Perm., Olig., Mio.-Rec.

Schizamminidae: Non-septate, tubular to flaring; wall labyrinthic; Rec.

Pasammosphaeridae: Globular to irregular; Ord.-Rec.

Subfamilies : 2

Superfamily | **Family Characters and Distribution**

Saccammina

Saccamminidae: Globular to elongate; aperture single to multiple, rounded to slit-like; Ord.-Rec. e.g., *Saccammina.* Subfamilies: 3

Polysaccamminidae: Linear series of irregular rounded chambers; ap. terminal; Rec., near shore/brackish water.

Hemisphaeramminidae: One or more globular chambers; aperture single, indistinct; Ord.-Rec.

Subfamilies: 3

Diffusilinidae: Irregular, a mass of branching tubes; aperture on the top of pistules; Ord.-Rec.

Komokiacea:

Test a complex system of branching tubes; agglutinations of argillaceous Rec.,

Families: 2

Komokiidae: Bushy or aborescent; generally non-septate, Rec.

Baculellidae: Sparsely branching aborescent; muddy particles; abyssal openings appearance; Rec.

Hyperamminacea:

Proloculus followed by tubular or flaring stage; agglutinated; aperture Terminal; rounded; M. Ord.-Rec.

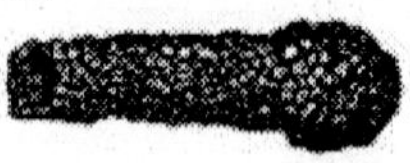

Hyperammina

Families: 3

Hyperamminidae: Elongate, tubular, flaring; Silu.-Rec.

Subfamily: 1

Hyperamminoididae: Elongate, flaring,; no true septa; U. Dev.-L. Cre.

Notodendrodidae: Large, central area bulbous; aperture indistinct; Rec.

Ammodiscacea:

Proloculus followed by tubular, enrolled nonseptate second chamber; aperture at the tube end; Agglutinated; L. Cam.-Rec.

Ammodiscus

Families: 1

Ammodiscidae: Proloculus globular; second chamber coiled or uncoiled; nonseptate; aperture at the end of tube; L. Cam.-Rec. e.g., *Ammodiscus*

Subfamilies: 5

Rzehakinacea:

Test enrolled; two to three chambers per whorl; coiled in a plane; or in varied plane, like milioline; finely agglutinated; Cre.-Rec.

Family: 1

Rzehakinidae: As in the superfamily; Cre.-Rec.

Superfamily | **Family Characters and Distribution**

Hormocinacea:

Multilocular, uniserial; agglutinated; L. Car.-Rec.

Rheophax

Families: 5

Aschemonellidae: Serial, inflated chamber; thin; aperture at the end of tube; L. Cret.-Rec.

Subfamilies: 2

Hormosinidae: Uniserial, may be curved; aperture terminal, single or multiple; L. Carb.-Rec. *Reophax*

Subfamily: 4

Thomasinellidae: uniserial, may produce 2 or more linear series; aperture terminal, slit-shaped; Cre.-Rec.

Dusenburyinidae: Elongate, numerous uniserial chambers; aperture terminal, with tooth; Rec.

Cribratinidae: Uniserial, straight; thick with subdermal alveolar layer; aperture terminal; L. Cre.-U. Cre.

Lituolacea:

Multilocular, planispiral to streptospiral coiling, may uncoil; agglutinated; non-perforate; U. Dev.-Rec.

Haplophragmoides

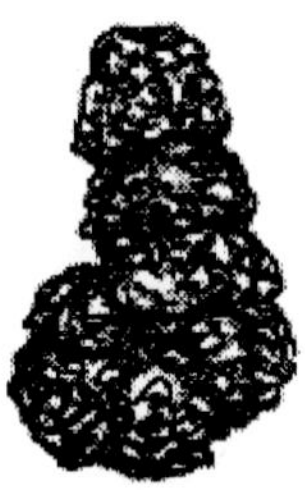

Ammobaculites

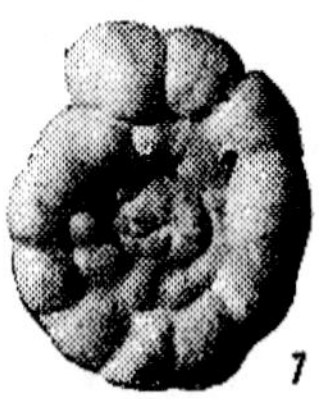

Trochamminoides

Lituola

Lituola

Families: 11

Oxinoxisidae: loose coil of subglobular chambers, then becoming straight; aperture terminal; L. Dev.-L. Carbo.

Haplophragmoididae: Planispirally coiled, involute; wall non- alveolar; aperture single, multiple, equatorial; Tri.-Rec. e.g., *Haplophragmoides*

Discamminidae: Planispirally coiled, evolute, may be uncoil, or straight; interiomarginal equatorial aperture; .Cre.-Rec.

Sphaeramminidae: Planispiral, evolute, later involute; finely agglutinated, aperture areal, rounded or slit; Rec. *Trochamminoides*

Lituotubidae: Proloculus followed by enrolled, elongate, chambers, later uncoil; aperture terminal; Cre.-Rec.

Lituolidae: Enrolled may be uncoil or straight in later stage; aperture terminal; L. Car.-Rec. e.g. *Lituola, Ammobaculites*

Subfamily: 4

Ammoshaeroidinidae: Streptospirally enrolled; aperture interiomarginal or areal; Jura.-Rec. Subfamily: 2

Haplophragmiidae: Streptospiral, may be uncoil and straight; aperture single or multiple; Jura.-Cre. e.g.,

Nezzazatidae: Trochospiral or planispiral; calcareous, probably, agglutinated, imperforate; aperture may be interomarginal or areal, single or multiple; Cre.

Subfamilies: 2

Barkerinidae: Planispiral, involute, may uncoil; chambers numerous; with interseptal partitions; aperture areal/interiomarginal, single/multiple; L. Cre.-U. Cre.

Placopsilinidae: Coiled, uncoil in later; aperture terminal; single or multiple; Jura.-Rec.

Superfamily | **Family Characters and Distribution**

Coscinophragmatacea: Coiled, may uncoil or branching in later stage; agglutinated; pores large; Tri.-Rec.

Families: 4

Lituoliporidae: Coiled, may uncoil; wall traversed by pores; Jura.

Biokovinidae: Enrolled, may uncoil; coarsely perforate; aperture simple, single or multiple; Jura.

Haddoniidae: Coiled to uncoiled, may be branched; aperture terminal, simple to complex; Eoc.-Rec.

Coscinophragmatidae: Similar to the Haddoniidae but with labyrinthic wall; Tri.-Rec.

Loftusiacea:

Planispiral to trochospiral, may uncoil or flare or peneropliform; agglutinated; Jura.-Rec.

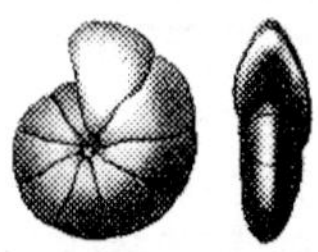

Clyclammina

Families: 3

Cyclamminidae: Enrolled, involute; Jura.-Rec. e.g., *Cyclammina* Subfamilies: 5

Spirocyclinidae: Trochospiral, may become planispiral and peneropliform; chambers subdivided; aperture cribrate; Jura.-Rec.

Loftusiidae: Large, planispiral; chambers subdivided; aperture multiple; Cre.

Spiroplectamminacea:

Planispiral to biserial; agglutinated; Carb.-Rec.

Family: 1

Spiroplectamminidae: Planispiral or streptospiral to biserial; Carb.-Rec. Subfamilies: 5

Trochamminacea:

Multilocular, low trochospiral; wall agglutinated; Crab.-Rec.

Trochammina

Families: 2

Trochamminidae: Trochospiral; Ap. interiomarginal to areal, single or multiple, may have supplementary umbilical openings; Carb.-Rec. e.g., *Trochammina*

Subfamilies: 8

Remaneicidae: Interior subdivided by secondary septa; Rec.

Subfamily: 2

Verneuilinacea:

Multilocular, tri- or biserial, later may become uniserial; agglutinated; Tris., Jura.-Rec.

Gaudryina

Tritaxia

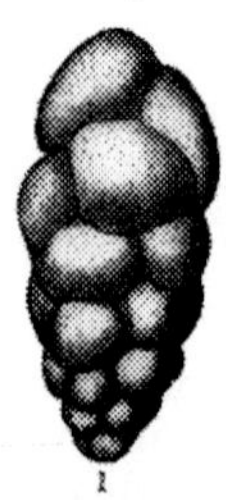

Verneuilina

Families: 5

Textulariopsidae: Bi- to uniserial, loosely bi- or uniserial; L. Jura.-U. Cre.

Verneuilinidae: Triser. may become biser., irregularly biserial/uniserial; ap. interiomarginal, may beslit-like, oval/ rounded/areal in biserial and uniserials; U. Tri.-Rec. e.g., *Verneuilina, Gaudryina, Tritaxia.*

Plectorecurvoididae: Biserial, with biserial axis Planispirally enrolled, as in calcareous Casiidulinidae; L. Cre.

Pseudobolivinidae: Biserial or loosely biserial; thin; aperture, oval or rounded areal, terminal or on neck, single or multiple; Rec.

Nouriidae: Biserial, loosely biserial; aperture terminal, ovate, may be on neck; Cre.-Rec.

Superfamily | **Family Characters and Distribution**

Tawitawiacea:

Large; biserial; interior divided by pillars; agglutinated; ap. high arched, at the base of ap. face, may become terminal and multiple; Oligo.-Rec.

Family: 1

Tawitawiidae: Biserial; Oligo.-Rec.

Ataxophragmiacea

Trochospiral, may reduce to biserial or uniserial or cyclic; chambers may have subdivisions; agglutinated; U. Tris., Jura.-Rec.

Dorothia

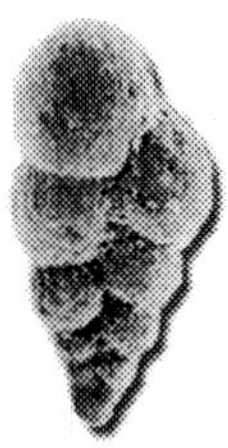

Heterohelix

Families; 9

Ataxophragmiidae: Trochospiral, 2-3 chambers per whorl may increase or decrease later; aperture, high, narrow arch; or straight to curved; interiomarginal or areal slit; Cre.-Paleo. Subfamilies: 2

Dorothiidae: Trochospiral, 4 or more chambers per whorl, may become bi- or uniserial; aperture interio-marginal or areal and rounded, without surrounding lip or neck; Jura.-Rec. e.g., *Dorothia* (old: *Marsonella*).

Subfamilies: 2

Eggerellidae: Trochospiral, at least earlier; tend to be tri-, bi- or uniserial; agglutinants generally calcareous; Ap. an interiomarginal slit to areal; Paleoc.-Rec. *Eggerella*

Subfamilies: 2

Globotextulariidae: High trochospirally coiled, 4 chambers per whorl; aperture interiomarginal; Rec.

Textulariellidae: Trochospiral, 3 or more chambers per whorl, tri-, bi-, or uniserial later; alveolar; Mio.-Rec.

Cuneolinidae: Conical, subflabelliform, trochospiral, 5 chambers per whorl, reduced to biserial; aperture multiple, at the base of apertural face; ?U. Tri.-Cret. Subfamilies: 2

Dicyclinidae: Discoidal, chambers cyclic & added biserially; subdivided; aperture multiple, peripheral; Cre. e.g., *Dicyclina*

Pfenderinidae: Trochospiral; chambers subdivided; alveolar wall; Jura.-Cre. Subfamily: 2

Coskinolinidae: Trochospiral, become uniserial, straight; aperture basal, cribrate; Jura.-Eoc

Textulariacea:

Trochospiral, bi- or triserial in early stages, later may be bi- or triserial; wall agglutinated; Cre.-Rec.

Textularia

Families: 3

Textulariidae: Biserial, at least in early stages, may become uniserial; aperture interiomarginal, to areal, single or multiple; Paleoc.-Rec., e.g., *Textularia* Subfamilies: 3

Superfamily	Family Characters and Distribution
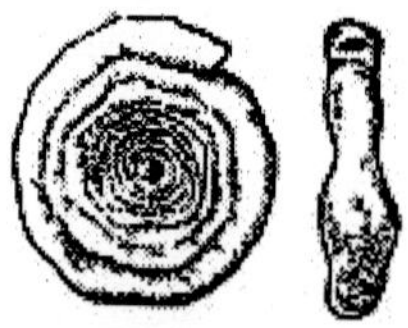 *Cyclolina*	**Valvulinidae**: Triserial, may become uniserial; aperture, with valvular tooth in early, multiple in later; Paleoc.-Rec. **Charysalidinidae**: Triserial, later biserial; chambers globular; aperture multiple; Cre.-Eoc. Subfamily: 1
Cyclolinacea: Discoidal, annular chambers in later stages, may be subdivided in advanced forms; agglutinated; L. Jura.- U. Cre.	Families: 2 **Cyclolinidae**: Discoidal, chambers annular in microspheric stages, may be subdivided; Cre. e.g., *Cyclolina* Subfamilies: 3 **Orbitopsellidae**: Coiled, later cyclic, stolon system radial; L. Jura.-U. Cre.
Orbitolinacea: Low conical; proloculous globular; chambers subdivided; agglutinated; Jura.-Eoc.  *Orbitolina*	Families: 2 **Dictyopsellidae**: Conical, proloculus globular, chambers divided by perpendicular beams; aperture interiomarginal; U. Cre. **Orbitolinidae**: Conical, early stage trochospiral, then straight; embryonic apparatus (in megalospheric) has proto-and deuteroconch; wall trilamellar, (upper calcitic); aperture with numerous pores; Jura.-Eoc. e.g., *Orbitolina* Subfamilies: 2

Suborder FUSULININA

Test wall homogeneously microgranular calcite of tightly packed equidimensional sungranular crystals, wall differentiated into two layers in advanced forms; Ord. -Tris.

Superfamily	Family Characters and Distribution
Parathuramminacea Unilocular; globular to elongate, or cluster of such chambers; Silu.-Tri.	Family: 1 **Parathuramminidae**: Aperture multiple at ends of tubular projections, or absent; Silu. -L. Carb.
Earlandiacea: Proloculus globular, second chamber undivided, straight or coiled; Silu. - Tris.	Families: 2 **Earlandiidae**: Test tubular non septate; Silu. - Tris. **Pseudoammodiscidae**: Proloculus followed by plani-spiral, trochospiral or streptospiral, undivided tubular second chamber; Dev. L. Carb.
Archaediscacea: Discoidal to globular, proloculus followed by undivided tubular chamber wall calcareous; L. Crab. -Perm.	Families: 2 **Archaediscidae**: Similar to the Superfamily. Subfamilies: 3 **Lasiodiscidae**: Discoidal to conical, two layered wall; aperture at the end of tube; L. Carb.-Perm.

Superfamily — **Family Characters and Distribution**

Moravamminacea:

Proloculus followed by tubular enrolled or straight chamber; Dev. L. Carb.

Families : 3

Caligellidae: Similar

Moravamminidae: Differs in having irregularly septate; wall microgranular; aperture terminal; M. Dev. - U. Dev.

Paratikhinellidae: Test interior partially subdivided; wall monolamellar, microgranular, calcite; aperture terminal; U. Dev. - L. Carb.

Nodosinellacea:

Wall with microgranular outer, and fibrous inner layer or single layered; Ord. - Per.

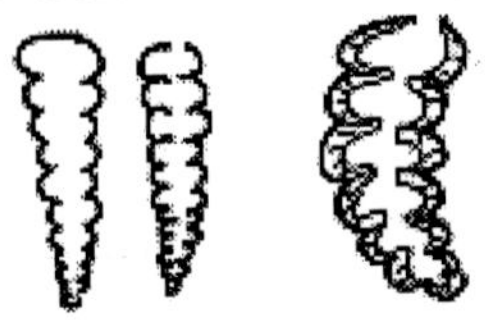

Earlanditina *Nodosinella*

Families: 3

Earlandinitidae: Elongate, uniserial, may be slightly curved; microgranular, single layer; aperture terminal; Ord.- U. Carb.

Tuberitinidae: Uni-, bilocular, subhemispherical, surface punctate; microgranular; aperture indistinct; Silu.-Perm.

Nodosinellidae: Uniserial; outer layer microgranular, inner fibrous; aperture terminal; U. Dev. - Perm.

Geinitzinacea:

Uniserial; rounded, ovate or fusiform in section; dark microgranular inner and radially fibrous outer layers; Dev. – Perm

Families: 2

Geinitzinidae: Early part rounded in section, later one compressed and ovate in section; outer layer light colored, hyaline radiate; aperture terminal; Dev. Perm.

Pachyphloiidae: Uniserial, compressed; microgranular calcite; L. Perm. - U. Per.

Colaniellacea:

Uniserial, chambers highly overlapping, internally, subdivided; outer layer; vitreous; aperture radiate; U. Dev. - Perm.

Family: 1

Colaniellidae: As in the superfamily

Ptychocladiacea:

Uniserial branching; microgranular calcareous; aperture indistinct; U.Carb.

Family: 1

Ptychocladiidae: As in the Superfamily

Paleotextulariacea:

Biserial, enrolled biserial, may become uniserial; wall microgranular calcareous; aperture single or multiple; in later stages; Dev. - Perm.

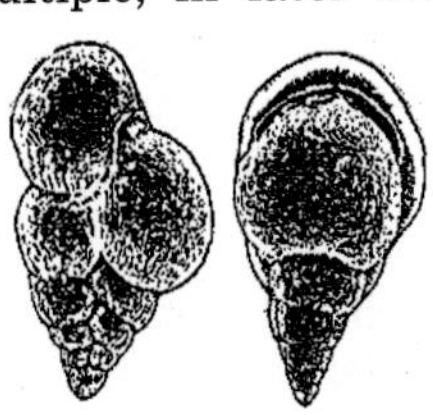

Pseudotextularia

Families: 3

Semitextulariidae: Biserial may become uniserial; microgranular with two layers; aperture basal or areal; single or multiple; Dev. - L. Carb.

Subfamilies: 3

Paleotextulariidae: Biserial, may become uniserial; microgranular; aperture interiomarginal; becoming terminal and cribrate in uniserial stage; Dev. - Perm.

Biseriamminidae: Biserial or biserialy Planispirally enrolled, may uncoil; microgranular; aperture at inner border of face; L. Carb. - Perm.

Subfamilies: 3

Superfamily	Family Characters and Distribution

Tournayellacea:

Proloculus followed by planispiral to streptospiral , enrolled second chamber, may uncoil; microgranular with a few agglutinations; aperture single or cribrate; U. Dev. - Perm.,

Families: 2

Tournayellidae: As in the superfamily.

Subfamilies: 5

Palaeospiroplectamminidae: Trochospiral, later planispiral and finally biserial; microgranular; aperture basal; L.Car.- L. Perm.

Endothyracea:

Planispiral, streptospiral may uncoil; microgranular; aperture single, basal or areal; Dev. - Tris.

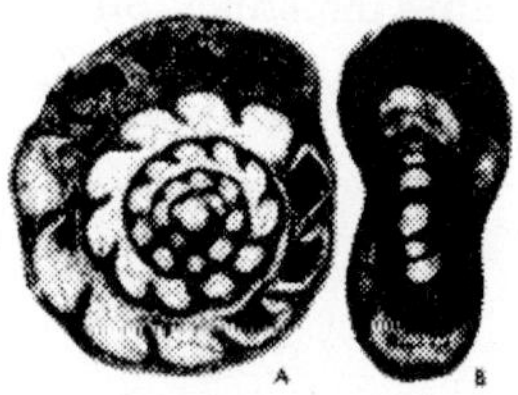

Endothyra

Families: 5

Endothyridae: Planispiral to streptospiral, involute, may uncoil; microgranular; Ap. simple, basal, areal or even cribrate; may have supplementary apertures; Dev. - Tris. Subfamilies: 4 e.g., *Endothyra*

Dariopsidae: Enrolled, may uncoil, septa incomplete; microgranular; L. Carb.

Bradyinidae: Planispiral, involute, rarely biserial; micro-granular with tectum/alveolar, tectum; Ap. simple, may cribrate, later; sutural openings; L. Carb.- Perm.

Loeblichiidae: Planispiral, throughout, evolute; chambers numerous; aperture basal; Dev. - Perm. Subfamilies: 2

Eocristellariidae: Enrolled, rapid increase in height; outer layer calcareous, hyaline; aperture basal, slit; Perm.

Tetratexacea:

Conical, trochospiral; chamber with chamberlets; microgranular; aperture umbilical; L. Carb. - Tris.

Families: 4

Pseudotaxidae: Wall single layered; Carboniferous.

Tetrataxidae: Wall two layered; L. Carb. to Tris.

Valvulinellidae: Conical, trochospiral; chambers subdivided; microcrystalline, calcareous; Ap. interiomarginal; Carb.

Abadehellidae: Conical, trochospiral; chambers subdivided; outer microgranular, inner fibrous; U. Perm.

Fusulinacea:

Enrolled, numerous cham/whorl; spherical/discoidal/fusiform; wall cal. , micro-granular of 4 distinct layers; tunnels & foramina, chomata, tectoria & axial fillings; U. Missi. (L. Carb.) - Perm.

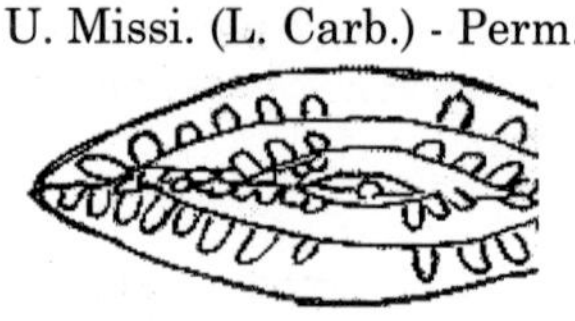

Fusulina

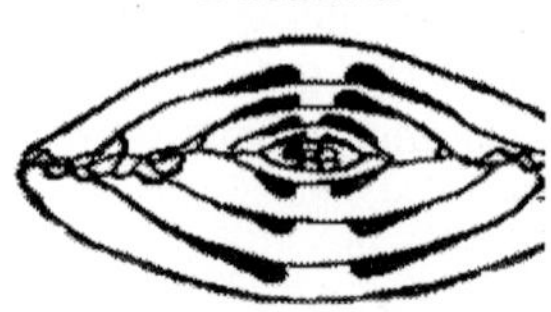

Profusulinella

Families: 8

Ozawainellidae: Discoid, spherical or ovoid; early forms planispiral evolute to irregularly coiled; wall (spirotheca) of early forms with tectum, late forms with diaphanotheca between tectum and lower tectorium; Missi. (L. Carb- Perm.)

Subfamilies: 2

Schubertellidae: Test fusiform to subcylindrical; advance taxa may be uncoiled and rectilinear; tunnel single, chomata low to large and asymmetrical; M. Pennsylvanian -U. Perm. Subfamily: 2

Fusulinidae: Fusiform to subcylindrical; planispirally coiled; septa flat to fluted; L. Pennsylvania – M. Perm . e.g. *Fusulina* Subfamily: 4

Superfamily

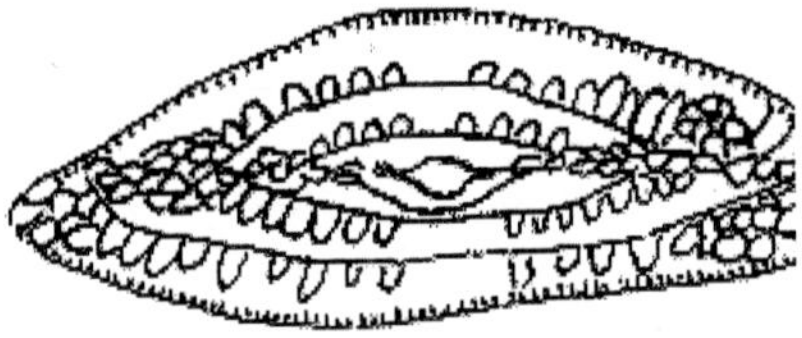

Schwagerina

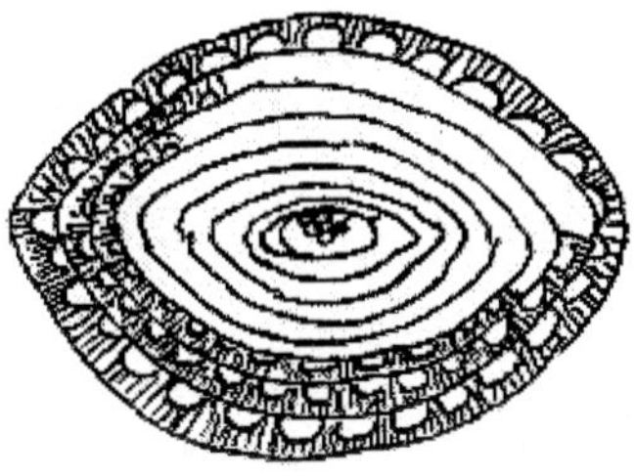

Neoschwagerina

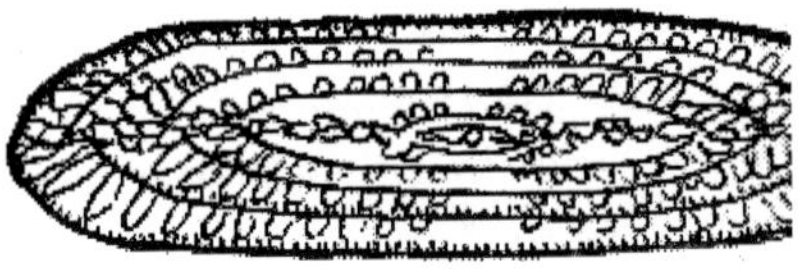

Parafusulina

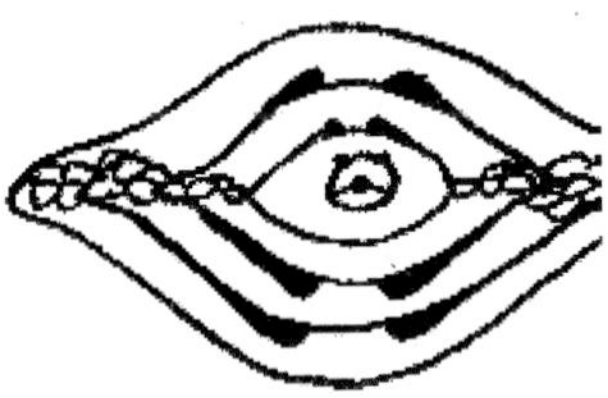

Fusulinella

Family Characters and Distribution

Schwagerinidae: Large, fusiform to irregularly cylindrical, planispiral and involute; septa strongly fluted; spirothica thick; M. Pennsylvania-Perm. e.g., *Profusulina*, *Schwagirina* Subfamilies: 4

Staffellidae: Small, subspherical to discoidal; septa closely spaced, simple; tunnel single; chomata distinct; wall two layered of tectum and diaphanotheca; U. Mississipian. - U. Perm.

Pseudoendothyridae: Discoidal; septa formed by the curvature of spirotheca; wall 4 layered, diaphanotheca thin; U. Miss. - L. Perm.

Verbeekinidae: Large, subspherical to cylindrical; planispiral, involute; spirotheca of tectum; L. Perm. - U. Perm.

Subfamilies: 5

Neoschwagerinidae: Fusiform to subcylindrical; spiro-theca consists of tectum and alveolar keriotheca; L. to U. Perm. e.g., *Neoschwagirina*

Subfamilies : 3

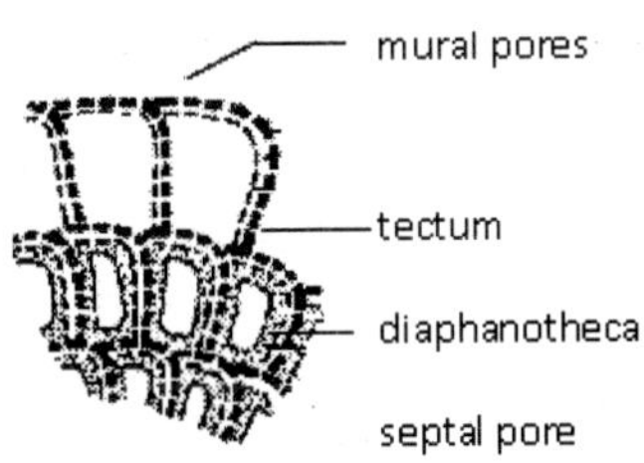

Fusulinid wall

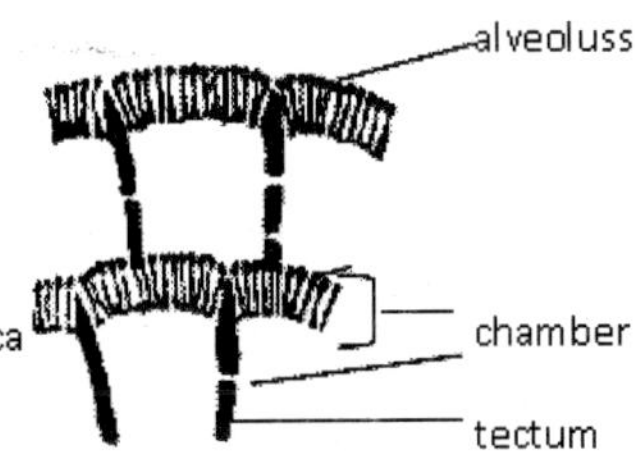

Schwagirinid wall

Suborder INVOLUTININA

Proloculus followed by enrolled tubular second chamber; wall calcareous, perforate radial, originally aragonitic (?) but may recrystallize to microgranular structure; pillar like structures at umbilicus; Tris.- Cre.

Superfamily

None

Involutina

Family Characters and Distribution

Families: 2

Involutinidae: Planispirally enrolled; L. Tris. - Cre.

Trocholinidae: Trochospirally coiled; Tris. - Cre.

Suborder MILIOLINA

Porcelaneous calcite, commonly with organic lining; post embryonic forms generally imperforate; spiral passage connecting proloculus and later chambers; Carb. - Rec.

Superfamily — **Family Characters and Distribution**

Squamulinacea:

Unilocular; imperforate, porcelaneous calcite; Rec.

Family: 1

Squamulinidae: As per superfamily

Cornuspiracea:

Planispiral or trochospiral; proloculus followed by undivided spiral chamber which may uncoil or become zigzag, and may be distinctly chambered; Carb.-Rec.

Families: 5

Conuspiridae: Second chamber undivided, planispiral to streptospiral, may become zigzag; Carb.- Rec. Subfam.: 5

Fischerinidae: Second chamber enrolled tubular with few chambers per whorl; Jura. - Rec. Subfamilies: 5

Nubeculariidae: Planispiral or irregularly coiled, at least in the early stage, later may branch out; aperture simple, rounded, slitlike or cribrate; Jura.-Rec. e.g., *Neobaeculinella*; Subfamilies: 3

Opthalmidiidae: Second chamber undivided, coiled chambers half a coil in length; U. Cre. - Rec.

Discospirinidae: Second chamber conuspirine coil, later chambers half a coil in length, and finally annular, and incompletely divided into chamberlets; M. Mio. - Rec.

Neobaeculinella

Miliolacea:

Two chambers/whorl arranged in varying planes about longitudinal axis; may uncoil; Jura. - Rec.

Families: 3

Miliolidae: Two chambers per whorl; chambers added in one to five or more planes; may very rarely uncoil; aperture at end of final chamber, simple or complex; Tris. - Rec.

Subfamilies: 6 (**Spiroloculininae**: chambers added at 180 degree apart e.g., *Spiroloculina, Pyrgo*; **Miliolinae**: chambers added at 142 degree, along five planes *Quiqueloculina*; **Hauerininae**; **Miliolinellinae**: 2 or 3 chambers visible externally e.g., *Triloculina*; **Sigmoilopsinae**; **Tubinellinae**)

Riveroinidae: Planispiral; chambers partially subdivided; aperture terminal; Rec.

Fabulariidae: Milioline at early stage; chambers subdivided; aperture multiple or single with tooth; U. Cret.-Eoc.? Olig.

Triloculina

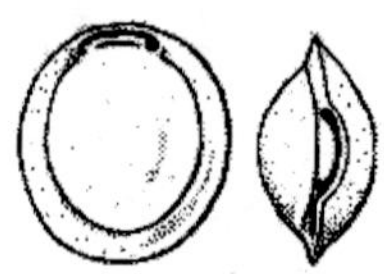

Pyrgo

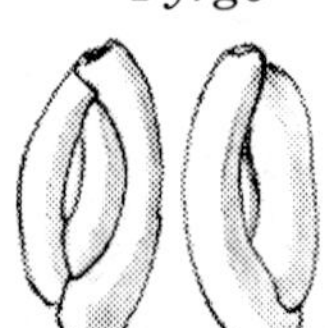

Quinqueloculina

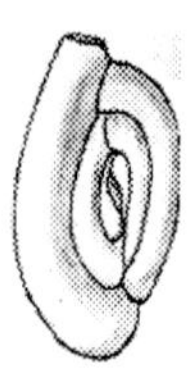

Spiroloculina

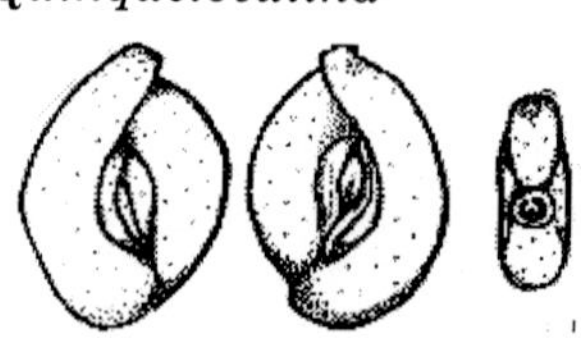

Spiroloculina

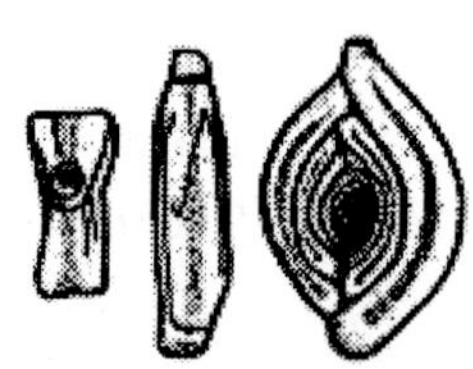

Spiroloculina

Haurina

Superfamily	**Family Characters and Distribution**
Soritacea: Chambers planispiral, uncoiling, flabelliform or cyclic, may be subdivided by interseptal partitions pillars; ? Perm.,Tris.-Rec	Families: 7
	Milioliporidae: Proloculus followed by chambers arranged in various planes; aperture terminal; ? Perm., Tris.
	Subfamilies: 3
	Peneroplidae: Close coiled in early, later uncoil or annular; or chambers not subdivided; aperture rounded; U. Tris., U. Cre.-Rec. e.g., *Peneroplis*
	Meandropsinidae: Planispiral, may uncoil, flare, flabelliform, cylindrical, conical or discoidal; aperture multiple; U. Cre.- M. Paleoc.
	Rhapydioninidae: Planispiral, or streptospiral, may uncoil; embryonic apparatus simple or quinueloculine; Jura., U. Cre. – M. Eoc., Rec.
	Subfamilies: 2
	Soritidae: Planispiral, may uncoil, flaring, fusiform or cylindrical; chamber numerous, subdivided; aperture multiple; U. Cre. - Rec. e.g., *Soritis*
	Subfamilies: 4
	Keramosphaeridae: Globular, concentric, chambers subdivided; U. Cret. - Paleoc., Mio., Rec.
	Alveolinidae: Large, globular, fusifrom or subcylindrical; chambers may be quinqueloculine; apertures numerous; L. Cret. – Rec. e.g., *Alveolina*

Peneroplis

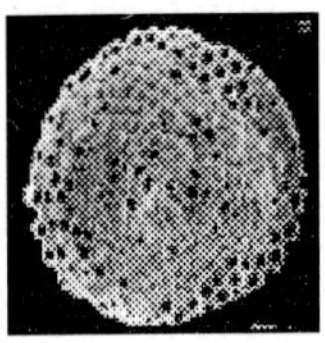

Soritis

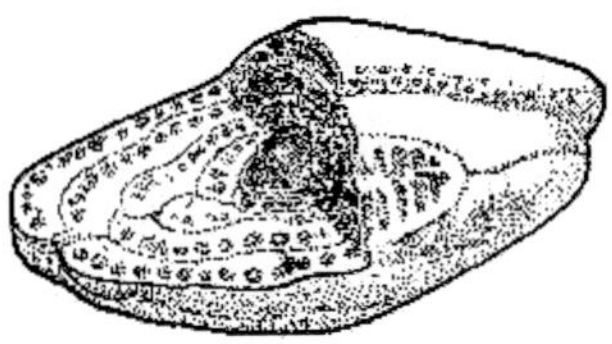

Alveolina

Suborder SILICOLOCULINA

Wall imperforate, of secreted opaline silica M. Mio. - Rec.

Superfamily	**Family Characters and Distribution**
Superfamily: None	Family: 1
	Silicoloculinidae: Small, coiled, proloculus followed by chambers of half- a-coil in length; plane of coiling changing as milioline; aperture terminal, arch slit; M. Mio. -Rec.

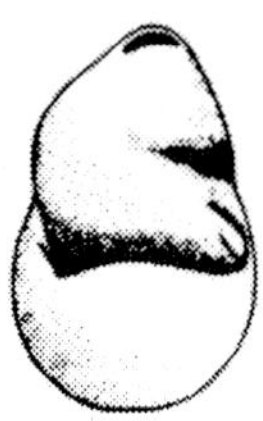

Silicoloculina

Suborder SPIRILLININA

Proloculus followed by enrolled tubular undivided-chamber or by few chambers per whorl; planispiral to high trochospiral; chambers may be secondarily subdivided; wall calcite, optically a single crystal or a few to a mosaic of crystals; a-axis along the axis of coiling? Tris., Jura. - Rec.

Superfamily

Superfamily: None

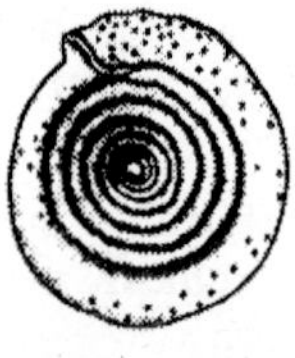

Sprillina

Patellina

Family Characters and Distribution

Families : 3

Spirillinidae: Proloculus followed by non-septate, enrolled tubular chamber, planispiral or trochospiral; aperture single, at open end. Tris., Jura. - Rec., *Sprillina*

Patellinidae: Proloculus followed by spiral, undivided, tubular chamber, later two chambers per whorl; L. Cre. - Rec. e.g., *Patellina*. Subfamilies: 2

Placentulinidae: Low trochospiral, non-septate, tubular, Undivided second chamber; few chambers per whorl; L. Jura. – L. Cre., Rec.

Subfamilies : 2

Suborder LAGENINA

Wall monolamellar, optically and ultrastructurally radiate calcite (c-axes of crystals perpendicular to surface); advanced taxa secondarily lamellar; M. Carb. - Rec.

Superfamily

Robuloidacea

Lagenina without secondarily lamellar wall; M. Carb.- L. Jura.

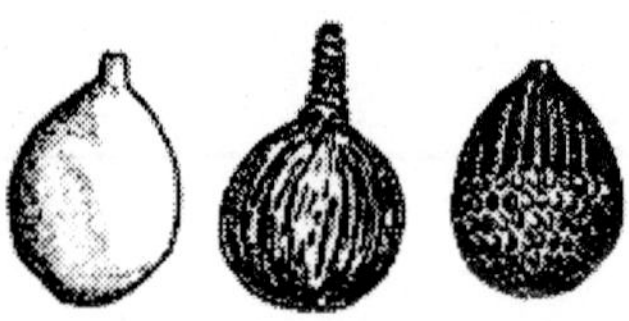

Legena

Family Characters and Distribution

Families: 3

Syzraniidae: Uniserial; wall single layered; M. Carb.-L. Jura.

Robuloididae: Uniserial, coiled, may uncoil or straight; Permian. e.g., *Lagena*

Partisaniidae: Biserial, later chambers added at 180 degree apart (known as sigmoidal arrangement), later may become uniserial ; aperture terminal; U. Permian.

Nodosariacea:

Wall with primary and secondary lamination as a result of continued growth; Tris. - Rec.

Families: 4

Nodosariidae: Multilocular, unilocular, biserial, straight; calcareous, hyaline, finely perforate, monolamellar; aperture terminal, radiate or multiple; Jura.-Rec.

Subfamilies: 3 (**Nodosariinae**: e.g., *Lagena, Nodosaria, Dentalina, Frondicularia)*

Nodogerina

Nodosaria

Dentalina

Frondicularia

Superfamily

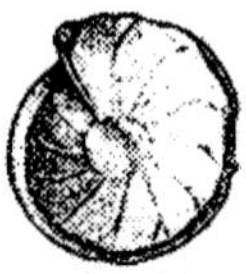

Lenticulina

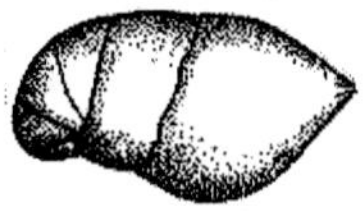

Marginulina

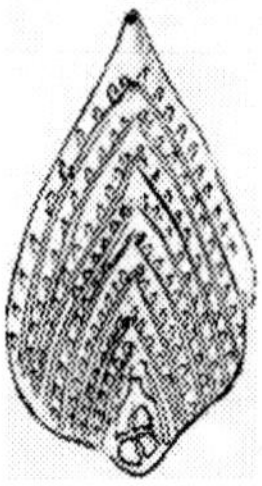

Neoflabellina

Polymorphina

Glandulina

Family Characters and Distribution

Vaginulinidae: Enrolled, early partially enrolled or arcuate, later uncoil and straight; aperture terminal, radiate or cribrate; Jura. - Rec.

Subfamilies: 4 (**Lenticulininae**: e.g., *Lenticulina;* **Marginulininae**: e.g., *Marginulina*; **Vaginulininae**: e.g., *Vaginulina, Neoflabellina.*)

Polymorphinidae: Spiral; chambers strongly overlapping towards the early part of the test; aperture at the distal end, radiate; Tris.- Rec.

Subfamilies: 3 **(Polymorphininae** e.g., *Polymorphina*)

Glandulinidae: Unilocular or uni-, biserial, polymo-rphine; aperture terminal, radial or slitlike; Tris., Jura. – Rec. e.g., *Glandulina*

Subfamilies: 4

Suborder ROBERTININA

Planispiral to trochospiral, internal partitions in chambers common; hyaline, perforate, radiate aragonite (c-axis normal to surface; very fine perforation; M. Tris. - Rec.

Superfamily

Duostominacea:

Planispiral to high trochospiral, non-lamellar, aragonitic which is granular; aperture single, double, interiomarginal; M. Tris. - L. Jura.

Robertinacea:

Basically trochospiral; chambers divided; aperture interiomarginal; Jura. - Rec.

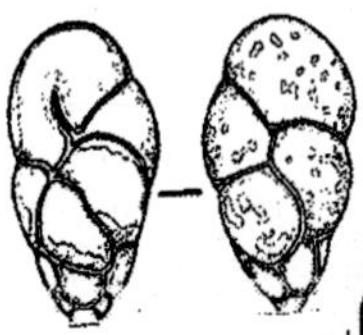

Robertina

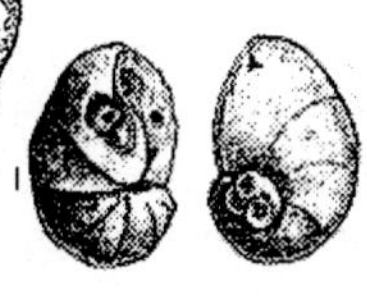

Ceratobuliminia

Family Characters and Distribution

Families: 3

Duostominidae: Spired; two interiomarginal; M. Tris.

Asymmetrinidae: Nearly planispiral, may be asymmetrical; U. Trias.

Oberhauserellidae: Low trochospiral; chambers some-what inflated dorsally; ap. interiomarginal slit; M. Tris.- L. Jura.

Families: 5

Conorboididae: Trochospiral, chambers per whorl may reduce finally to uniserial; ap. interiomarginal umbilical in trochospiral; terminal in uniserial; Jura. – M. Eoc.

Ceratobuliminidae: Trochospiral; aperture interio-marginal; chambers subdivided; L. Jura. - Rec.

Subfamilies: 2

Robertinidae: High trochospiral; L. Eoc.-Rec. e.g., *Robertina, Ceratobulimina* (Subfamilies: 4) Other families:

Epistominidae: Jura.-Rec. Subfamilies 2

Mississippinidae: Eoc. Rec.

Suborder GLOBIGERININA

Planktonic, perforate hyaline calcite, radiate; primary lamination bilamellar; L. Jura. Rec.

Superfamily

Heterohelicacea:

Biserial, triserial, later reduced to uniserial; aperture basal, arch or terminal in uniserial; L. Jura. Rec.

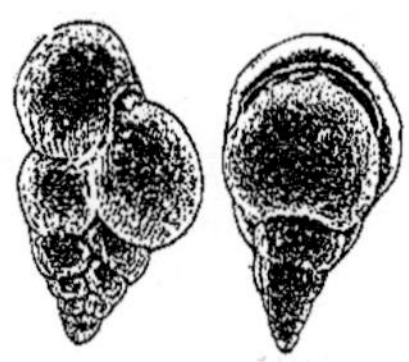

Heterohelix *Pseudotextularia*

Planomalinacea:

Planispiral may become trocho.; ap. equatorial & interio-marginal, with lip; L. Cre. - U. Cre.

Planomalina

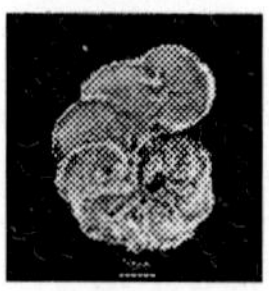

Rotaliporacea: Families:6

Trocho.; pri. ap. extra- umbilical, with lip; sutural openings in advanced taxa; M. Jura. - U.Cre.

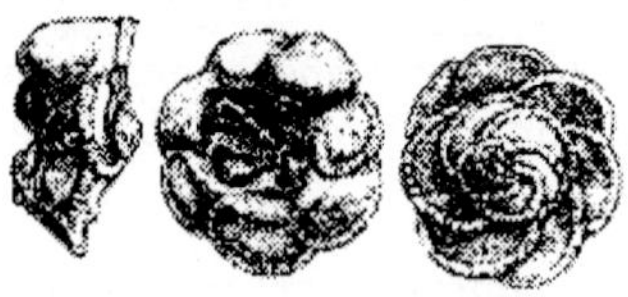

Globotruncana

Globotruncannacea: Families: 2

Trochospiral, globular toangular chambers, U. Cre.

Hantkeninacea: Family: 3

Planis. /enrolled biserial; chambers globular to elongate; aperture equatorial, single or multiple, may have relict or areal secondary apertures; Paleoc. -Mio., ?Plio

Family Characters and Distribution

Families: 2

Hetrohelicidae: Aperture symmetrical, slit to high arch at the base or terminal; L. Jura. - Oligo., Rec.? Subfamilies: 4 (Heterohelicinae: *Heterohelix, Pseudotextularia, Glublerina*

Chiloguembelinidae: Biserial, somewhat twisted; apertures asymmetrical, up to the face of final chamber; Paleo.- Pleis.

Gublerina

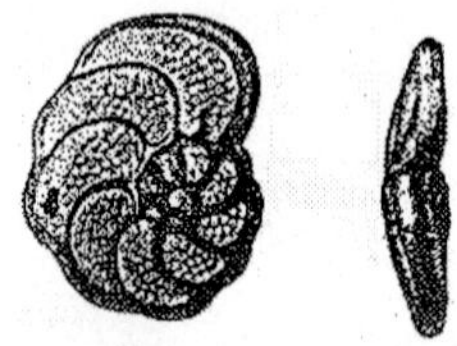

Planulina

Families: 3

Globigerinelloididae: Planispiral, globular; ap. basal, equa. & interiomarginal, with lip; L. Cre. - U. Cre. Subfamilies: 2

Planomalinidae: Planispiral, biumbili.; peri. carinate sutures strongly arched; ap. an equatorial opening; Cre.

Schackoinidae: Trochospiral-nearly planispiral; chambers with hollow spines; ap. equatorial, may have broad lip; Cre.

Hedbergellidae: Trochospiral; Ap. primary; lip of earlier chambers visible around umbilicus; M. Jura.-U. Cre. Subfamilies: 2

Globuligerinidae: Trochoid, few chambers per whorl; wall perforate; aperture umbilical, arch, with lip; Jura.-L. Cre.

Favusellidae: Trochospiral; chambers inflated; wall cancellate; ap. umbilical, lip; umbilicus deep; margin non-carinate; L. Cre.-U. Cre.

Rotaliporidae: Trochospiral; aperture umbi- extraumbilical, lip; sutural aperture on umbilical side; Cre. Subfamilies: 2

Globotruncanidae: Low Trocho.; cham. angular; peri. Truncate/carinate; ap. Umb., covered by tegilla with ap. s; U. Ceno., Maas. Peri. carinal-bands e.g., *Globotruncana*).

Rugoglobigerinidae: Trocho. inflated; peri. not carinated; aperture umbilical with tegilla; U. Cre. Tur.- Mass.

Globanomalinidae: Planispiral, slightly asymm. Biumbili. Cham. inflated; surface non-spinose; Paleoc.-Oligo., ? Plio.

Hentkaninidae: Planispiral; chamb. rounded with peri. spines; aperture interiomarginal and equatorial; U. Eoc.

Cassigerinellidae: Planis.-biserially enrolled; Ap. Equa. in planispiral, extraumbilical in biserial; Oligo.-Mio.

Superfamily | **Family Characters and Distribution**

Globigerinacea:

Trochospiral; wall perforate, surface with monocrystalline spines; interiomarginal, umbilical or equatorial, large sutural apertures may occur; Eoc-Rec.

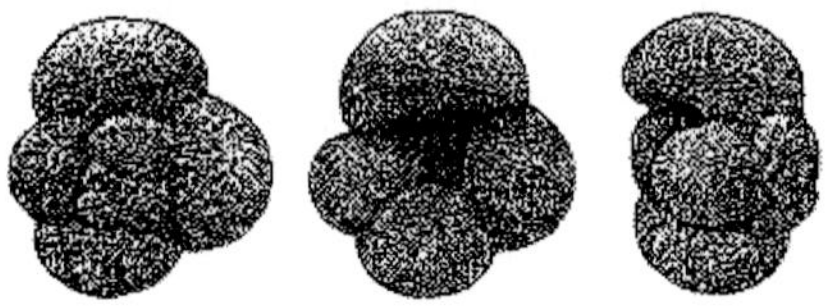

Globigerina

Families: 3

Globigerinidae: Trochospiral may become planispiral; highly perforate, aperture interiomarginal, umbilical or equatorial; sutural openings on dorsal side may be found; Eoc.-Rec. Subfamilies: 2 (**Globigerininae**: Trochospiral throughout; e.g., *Globigerina.*

Candeinidae: Trochospiral, early stage with umbilical aperture and spinose wall; umbilicus closed; apertures small rounded openings along sutures; Mio.-Rec.

Hastigerinidae: Trochospiral, later strepto- or planispiral; chambers globular to clavate; surface spinose; Pleis.-Rec. e.g. *Hastigerinoides*

Globorotaliacea:

Trochospiral; periphery rounded to carinate; wall finely lamellar, radiate, calcite; surface smooth, pitted, nonspinose; secondary apertures with bullae; U. Cre.-Rec.

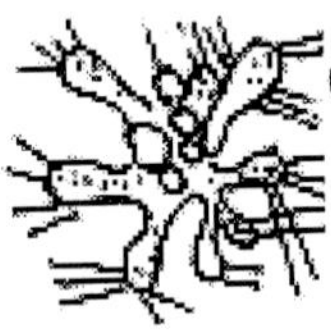

Hastigerinoides

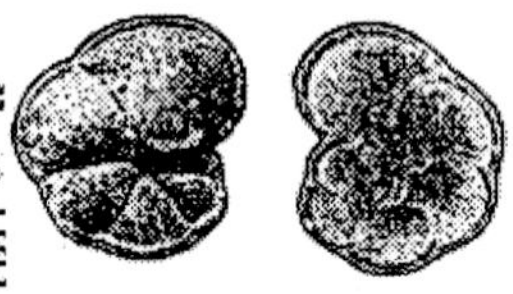

Globorotalia

Families: 5

Eoglobigerinidae: Low/moderate trochospiral; small opening along umbilicus; wall smooth, non-spinose, finely perforate; aperture, extra-umbilical, interiomarginal; Paleoc .-Rec.

Globorotaliidae: Trochospiral, subglobular to lenticular; surface smooth; aperture with thick lip; U.Cre.-Rec. e.g., *Globorotalia*

Subfamilies: 2

Pulleniatinidae: Streptospiral; chambers subglobular; perforate; surface non-spinose; U. Mio.-Rec.

Globigerinitidae: Trochospiral; surface smooth or pustulose; M. Eoc.-Rec,

Subfamilies: 2

Catapsydracidae: Trochospiral; pitted to cancellate but non-spinose; Paleoc.-Rec.

Suborder ROTALIINA

Multilocular, enrolled may become biserial or uniserial; chambers simple or subdivided; wall calcareous, perforate hyaline, lamellar calcite; formed by calcification at each side of organic membrane; aperture simple or with internal tooth plate; internal canal sys. resent in some; Tris.-Rec.

Superfamily | **Family Characters and Distribution**

Turrilinacea

Trochospiral or triserial, may become bi- or uniserial; calcareous, perforate, radial; aperture simple, no tooth plate; U. Cre.-Rec.

Turrilina

Families: 4

Turrilinidae: Trochospiral, 3 or more chambers per whorl, may become biserial; Eoc.-Rec.

Trimosinidae: Tri- may reduced to biserial; spines along chambers; aperture large, interiomarginal; Rec.

Pavoninidae: Biserial, later chambers broad, arch shaped or flabelliform; aperture terminal; Eoc.-Rec.

Sphaeroidinidae: Streptospiral, chambers strongly embracing; aperture an arch slit, above base of chamber; suppl. openings present; U. Cre.-Rec.

Superfamily	Family Characters and Distribution

Eouvigerinacea:

Bi- or uniserial, may have initial planispiral coil biserial stage may be enrolled; perforate, optically radiate; Jura.- Rec,

Families: 7

Bolivinidae: Bi- later may be uniserial; aperture high loop shaped; Jura.-Rec. e.g., *Bolivina*

Bolivinoididae: Biserial, calcareous, perforate, radiate; high interiomarginal aperture; U. Cre.-Paleoc.

Islandiellidae: Biserial or biseriality enrolled planispiral coil; calcite, radiate; aperture interiomarginal: Paleoc.-Rec. Subfamilies: 3

Elhasaellidae: Planispiral (/), later tri-, bi- and finally uniserial; surface spinose; aperture terminal; U. Cre.

Eouvigerinidae: L. Cret. – M. Eoc.

Stilostomellidae: Eoc.-Rec.

Lacosteinidae: Planispiral-biserial or up to 4 chambers/ whorl; ap. High- interiomarginal; U. Eoc.- ? Plio.

Bolivina

Buliminacea:

Trochospiral, may become tri-, bi- or even uniserial; calcareous perforate, radial; Aperture high loop shaped interiomarginal opening; M. Jura.-Rec.

Families: 9

Praebuliminidae: Aperture an arch, may become terminal or may cribrate; M. Jura. -Paleoc.

Bolivinitidae: Biserial, flattened, angles carinate; aperture a high opening on face of last chamber; Mio.-Rec.

Stainforthiidae: Biserial, later twisted biserial; chambers high; aperture loop shaped on apertural face; Eoc.-Rec.

Buliminidae: Trochospiral, less than 3 chambers per whorl, may become biserial; aperture loop, on the face or areal pores; Paleoc.-Rec. e.g., *Bulimina, Bolivinoides*

Buliminellidae: High trochospiral, many chambers per whorl; aperture loop-shaped; ? U. Cre., Paleoc.-Rec.

Reussellidae: Triserial, late bi- or uniserial; aperture interiomarginal/ terminal in uniserial; Eoc.-Rec., *Reusella*

Siphogenerinoididae: Tri-, bi- or uniserial; aperture with toothplate; U. Cre.-Rec.

Subfamilies: 2

Uvigerinidae: Tri- to biserial, intend to be bi- uniserial; aperture terminal with neck and phialine lip; and provided with internal siphon like toothplate connecting apertures of successive chambers; U. Cre.-Rec.

Subfamilies: 2 (**Uvigerininae**: Chambers rounded or inflated; aperture terminal with neck and phialine lip e.g., *Uvigerina*)

Millettiidae: Biserial, later uniserial; subdivided chambers; aperture terminal, lip phialine; Rec.

Bulimina

Bolivinoides

Reusella

Uvigerina

Superfamily	Family Characters and Distribution
Fursenkoinacea: Triserial, twisted or flattened biserial; wall calcareous, perforate, granular; optically hyaline oblique; aperture loop shaped; U. Cre.-Rec.	Families: 2 **Fursenkoinidae**: Twisted biserial, later uniserial; aperture loop shaped or terminal in bi-/uniserial stage; U. Cre.-Rec. Subfamilies: 2 **Fursenkoininae**: Biserial or twisted biserial; U. Cre.-Rec. **Sigmavirgulininae**: Chambers added at less than 180 degree in biserial test, resulting in sigmoid arrangement; Mio.-Rec. e.g., *Sigmavirgulina*. **Virgulinellidae**: Tri- to biserial; aperture loop shaped, supplementary sutural aperture; Mio.-Plio.

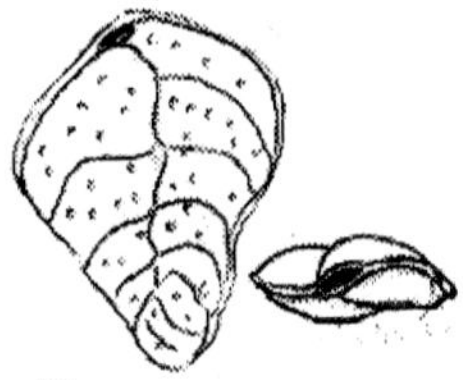

Sigmavirgulina

Superfamily	Family Characters and Distribution
Cassidulinacea: Enrolled biserial, later uncoiled to uniserial; calcareous, perforate; aperture interiomarginal slit or may become terminal; U. Cre.-Rec.	Families: 4 **Cassidulinidae**: Biserial, with biserial axis planispirally enrolled, late may uncoil; Eoc.-Rec. Subfamily: 2 **Loxostomatidae**: Biserial, may become uniserial; aperture low interiomarginal opening in biserial terminal slit uniserial; U. Cre.-Eoc. e.g., *Loxostomum*. **Bolivinellidae**: Biserial, chambers increasing rapidly to become palmate; aperture cribrate; Eoc.-Rec. **Annulopatellinidae**: Conical, second chamber reniform, than test becomes uniserial appearing annular; chambers subdivided; Mio.-Rec.

Loxostomum

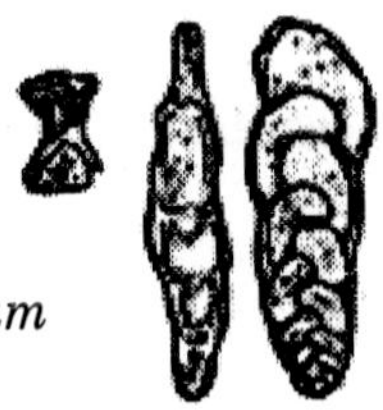

Superfamily	Family Characters and Distribution
Pleurostomellacea: Tri- or biserial, later uniserial; calcareous, perforate, lamellar, granular; aperture subterminal straight to curved slit, or terminal cribrate; ? Jura., L. Cre.-Rec.	Family: 1 **Pleurostomellidae**: As in superfamily. Subfamilies: 3
Delosinacea: Trochospiral or tri-, bi- or uniserial; perforate, hyaline; surface smooth; aperture interiomarginal slit, may be replaced by sutural pores; L. Cre.-Rec.	Families: 3 **Caucasinidae**: Trochospiral at least in early stage L. Cre.-Rec. Subfamilies: 2 **Tremachoridae**: Low trochospiral; aperture slit; openings at sutures; Mio. **Delosinidae**: Triserial, high spired; only sutural pores; Rec.

Discorbacea:

Low trochospiral; perforate, hyaline, radiate calcite; aperture interiomarginal; U. Cre.-Rec.

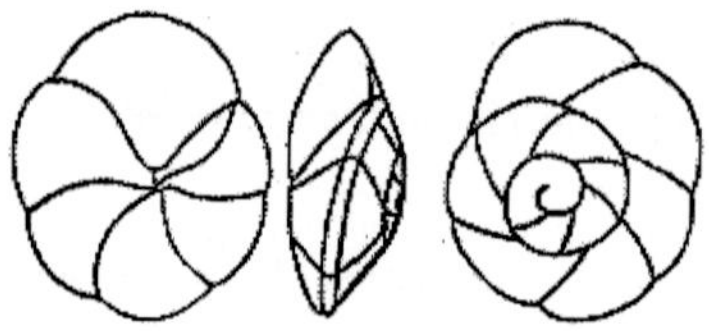

Eponides

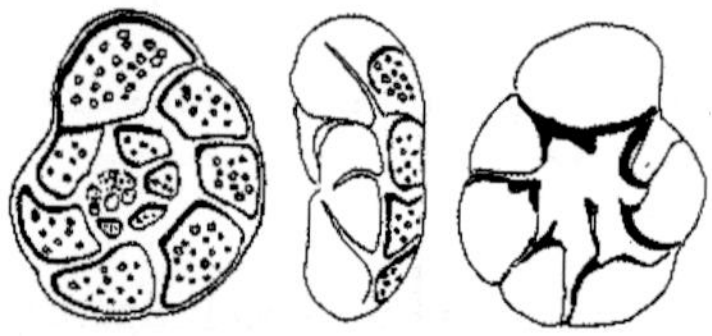

Rotorboides granulosum

Families: 9

Conorbinidae: Trochospiral; periphery angled; chambers lunate, last one covering up to one half of periphery; Cre.

Bagginidae: Trochospiral; finely perforated; Cre.-Rec. Subfamilies: 2

Eponididae: Trochospiral, at least early part; aperture interiomarginal or cribrate and areal; Paleoc.-Rec. Subfamilies: 4 (Eponidinae: relatively large interio-marginal aperture e.g., *Eponides,*

Bueningiidae: Enrolled, dorsal side inflated, ventral flattened, both involute, with keel and deep umbilicus; aperture umbilical, with small lip; L. Mio.-Plio.

Pegidiidae: Trochospiral, chambers few, early chambers resorbed with growth; ap. in series on umb. wall; Mio.-Rec.

Discorbidae: Trochospiral, umbilical region open; aperture umbilical commonly with projecting flap or lip; Eoc.-Rec. e.g., *Rotorboides granulosum, Discorbis*

Subfamilies: 2 **Discorbinae**: Low trocho.; ap. with distinct flap. **Rosalininae**: trocho., may have umbonal boss; ap. arch at base of final chamber on the ventral side; Mio.-Rec. **Pannellainidae**, **Bronnimanniidae** and **Rotaliellidae.**

Glabratellacea:

Trochospiral, hyaline, radial, calcite, perforate; radial striae on dorsal side; aperture interiomarginal; Eoc.-Rec.

Families: 2

Glabratellidae: Low trochospiral, radial ornamentation on dorsal side; Eoc.-Rec. e.g., *Glabratella*

Buliminoididae: Extremely high trochospiral; aperture umbilical in early portion; Olig. -Rec.

Siphoninacea:

Trochospiral; aperture slit-like; oval in apertural face/on neck, phialine lip; U. Cre.-Rec.

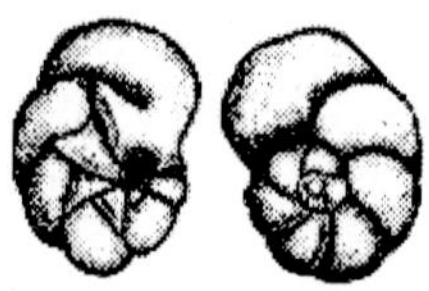

Discorbis

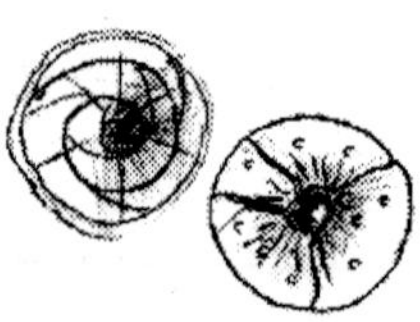

Glabratella

Families: 4

Siphoninidae: Trochospiral, at least early part; later planispiral uncoil, biserial; aperture rounded/oval with lip; U Cre.-Rec. Subfamilies: 3

Pseudoparrellidae: Trochospiral, at least early part; aperture vertical slit in apertural face/ parallel to peripheral margin or areal; Eoc.-Rec.

Subfamilies: 3

Planulinoididae: Planispiral, periphery truncate, biconcave, partially evolute on both sides; aperture areal, oblique and peripheral with supplementary aperture on umbilical side at inner margin of chambers; Plio.-Rec.

Discorbinellidae: Planispiral or slightly trochospiral; test may carinate; supplementary openings besides a primary interiomarginal aperture; Paleo.-Rec.

Subfamilies: 2

Superfamily | **Family Characters and Distribution**

Planorbulinacea:

Trochospiral, at least early part, may uncoil or straight or biserial or many chambers /whorl; perforate, hyaline radial, calcite; U. Cre. -Rec.

Families: 5

Planulinidae: trochospiral to nearly planispiral, partially evolute; aperture interiomarginal equatorial arch; U.Cre.-Rec. Subfamilies: 2 (**Planulininae**: Sides flat e.g., *Planulina*)

Cibicididae: Trochospiral enrolled, at least early part, aperture low equatorial arch, may extend onto spiral side; Cre.-Rec.

Subfamilies: 3 (**Cibicidinae**: Test enrolled throughout e.g., *Cibicides*).

Planorbulinidae: Early Trochospiral, later discoid, cylindrical, conical; aperture single or multiple, peripheral; Eoc.-Rec.

Cymbaloporidae: Trochospiral, later in annular series; numerous apertures as small circular pores; Cre.-Rec.

Subfamilies: 3

Victoriellidae: Trochospiral, later erect; aperture slit; U. Cre. Rec. Subfamilies: 3

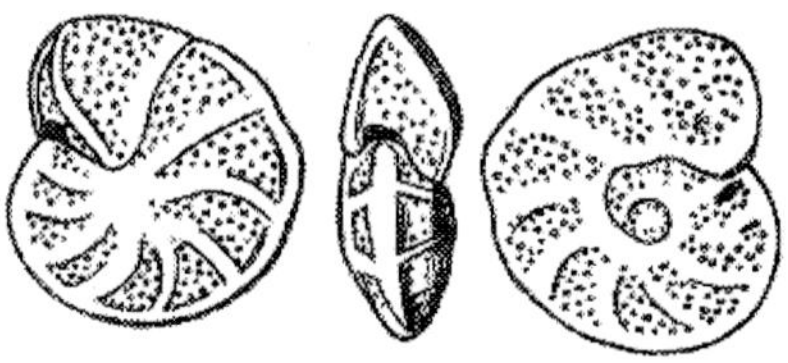

Cibicides

Acervulinacea:

Spiral, later irregular chambers; hyaline radiate, coarsely perforate calcite; no aperture, only mural pores; Eoc.-Rec.

Families: 2

Acervulinidae: Spiral stage followed by one/more layers of chambers; only mural pores; Eoc.-Rec.

Homotrematidae: Irregularly trochospiral, later chambers erect; aperture large, covered by porous plate; Eoc.-Rec.

Asterigerinacea:

Trochospiral to planispiral; chambers subdivided; no stolon system; perforate, radial, calcite; Cre.-Rec.

Families: 5

Epistomariidae: Trochospiral with supplementary chambers on ventral side; aperture, interiomarginal, supplementary sutural and areal apertures also present; Cre.-Rec.

Subfamilies: 4 (**Epistomariinae**: Supplementary openings on both sides e.g., *Epistomina*).

Alfredinidae: Supplementary apertures in umbilical region on ventral side; Eoc.-Rec.

Asterigerinatidae: Trochospiral, secondary chamberlets on ventral side; U. Cre.-Rec.

Asterigerinidae: Primary chambers alternating with secondary chamber which form stellate series around umbilical plug; Cre.-Rec. e.g., *Pseudoeponides*.

Amphisteginidae: Chambers numerous with complex chamberlets in center of umbilical side; aperture narrow, interiomarginal slit; Paleoc.-Rec. e.g., *Amphistegina*.

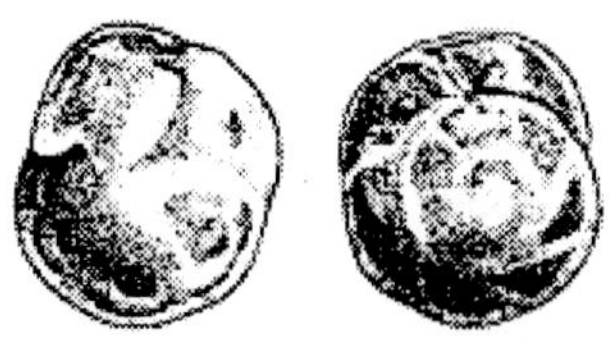

Epistomina

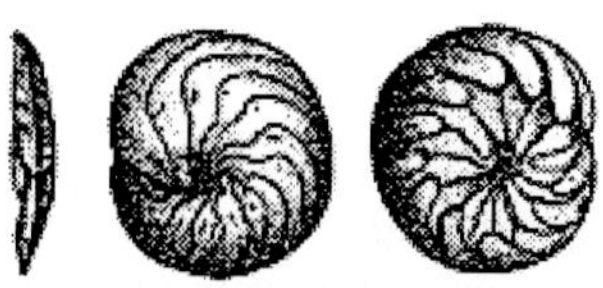

Amphistegina

Superfamily

Nonionacea:

Planispiral, slightly asymmetrical; perforate, granular calcite; aperture equatorial and interiomarginal; may have slitlike openings at periphery; U.Cre. - Rec.

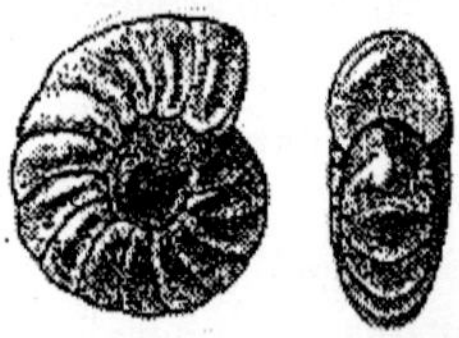

Nonion

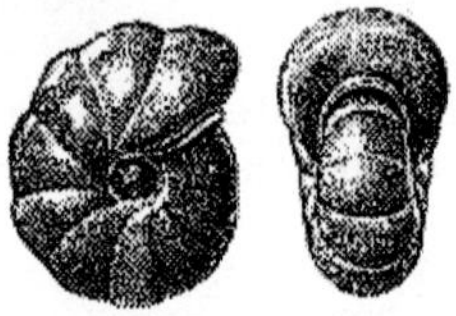

Melonis

Family Characters and Distribution

Families: 4

Nonionidae: Planispiral or nearly so; involute to evolute; aperture interiomarginal an equatorial; U.Cre.-Rec. e.g., *Nonion, Pullenia*

Subfamilies: 4

Melonidae: Planispiral, slightly asymmetrical; involute, biumbilicate; aperture arch shaped; ? U. Cre., Paleoc.-Rec. e.g., *Melonis*

Spirotectinidae: Trochospiral, dorsal side partially involute; sutures with septal bridges; aperture basal, equatorial; Rec.

Almaenidae: Nearly planispiral; primary aperture interiomarginal equatorial; secondary slitlike apertures at outer peripheral margins; Eoc.-Rec.

Subfamilies: 2

Chilostomellacea:

Enrolled, trochospiral, may uncoil later; perforate granular calcite; aperture extent from periphery to umbilicus on ventral side or up to apertural face; Jura. Rec.

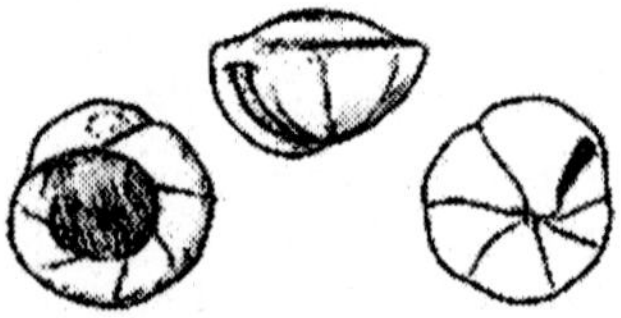

Alabamina

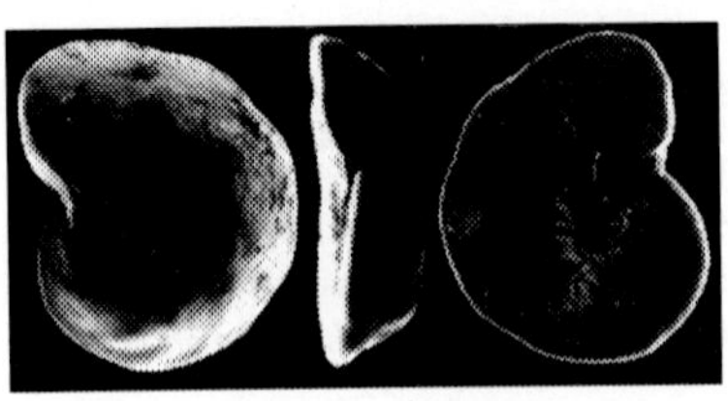

Hanzawaia

Families: 11

Quadrimorphinidae: Trochospiral; chambers not enveloping; L.Cre.-Rec.

Chilostomellidae: Trochospiral to planispiral; chambers much enveloping , those of outer whorl visible; Jura-Rec.

Globorotalidae: Trochospiral, planoconvex, pseudo-umbilicate; ventrally; aperture in between umbilicus and periphery; Cre.

Parrelloididae: Trochospiral, lenticular; umbilical boss. U. Cre.-Rec.

Alabaminidae:Trochospiral; aperture slit, interio-marginal, Cre.-Rec. e.g., *Alabamina* Subfamilies: 2

Oridorsalidae: Lenticular; finely perforate, optically granular; aperture basal, extending towards umbilicus; sutural openings at the junction of septal and spiral sutures, on dorsal side; Oligo.-Rec.

Osangulariidae: Aperture has two openings, interiomarginal, other vertical or oblique areal; L. Cre.-Rec.

Gavelinellidae: Trochospiral; Ap. interiomarginal, arch shaped, which extends towards umbilicus covered by flap; flaps generally visible in umbilical area; L. Cre.-Rec. e.g., *Hanzawaia*

Superfamily	Family Characters and Distribution
	Karreriidae: Trochospiral, later uncoil and straight; aperture terminal rounded; L. Cre.-Rec.
	Coleitidae: Trochospiral later uncoil; wall highly cancellate; Ap., elongate, terminal with projecting tooth; Paleoc.-L. Eoc.
	Trichohyalidae: Trochospiral; perforated vesicular plate obscure the umbilical cavity; Rec.
Orbitoidacea: Discoidal/lenticular; early portion has bilocular embryonic chambers; chambers subdivided; dimorphism prominent; megalospheric forms have distinct embryonic stage; wall perforate, optically radial calcite; openings between chambers present; replaced by stolon system in advanced forms; no canal system; U.Cre.-Mio.	Families: 3 **Linderinidae**: Megalospheric embryonic stage biloculine, later chambers in annular series; Eoc.-Mio. **Orbitoididae**: Embryonic chambers covered by perforated or by periembryonic chambers; stolons present; U.Cre. Subfamilies: 2 **Lepidocylinidae**: Discoidal to inflated lenticular; chambers walls perforated by stolons; M Eoc.-M. Mio. Subfamilies: 2
Rotaliacea: Trochospiral to planispiral; involute to evolute; many chambers per whorl; septal flaps, radial canals, fissures, umbilical cavities present; perforate hyaline calcite optically granular, primary aperture occur without any connection with interior of chambers; U. Cre. - Rec.	Families: 5 **Rotaliidae**: Trochospiral throughout, with radial canals or fissures or interseptal and subsutural canals; U. Cre.-Rec. Subfamilies: 5

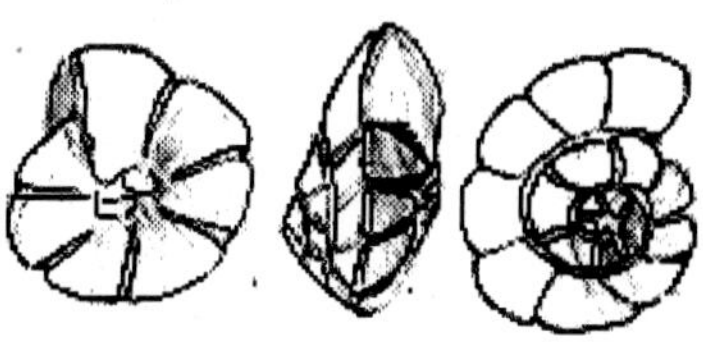

Cavarotalia

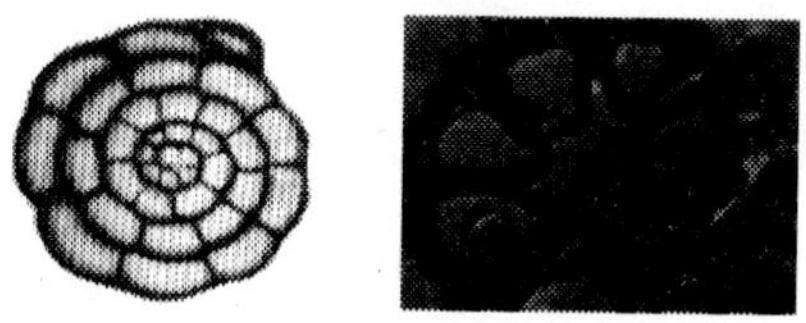

Cavarotalia

Superfamily | **Family Characters and Distribution**

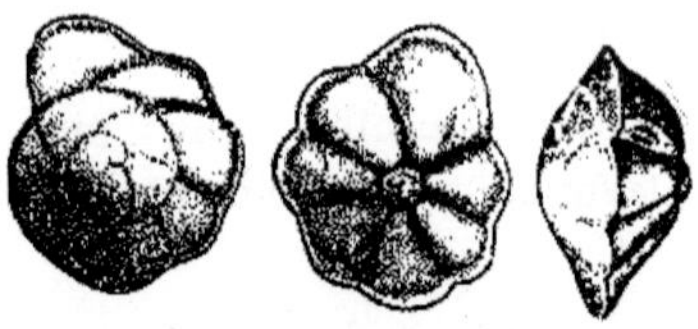

Pararotalia

Pararotaliinae: No canal system but with umbilical cavities single; interiomarginal slit-like aperture, converted into areal intercameral foramen by later attachment of tooth plate; U. Cre.-Rec. e.g., *Pararotalia.*

Rotaliinae: Trochospiral, external openings on ventral side, consisting of radial canals or fissure or umbilical cavities, commonly has intraseptal and subsutural canals; Upp. Cre.-Rec. *Cavarotalia, Ammonia, Pesudorotalia, Asterorotalia.*

Ammonia

Calcarinidae: Enrolled, small, dorsal and ventral surface undifferentiated; large spines common; canal system diffuse and confused with perforations; Upper. Cre.-Recent.

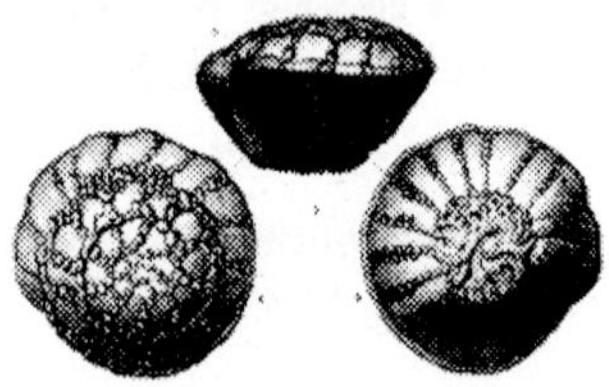

Pseudorotalia

Elphidiidae: Planispiral to trochospiral, may uncoil; sutural canal system opening into sutural pores; aperture interiomarginal or areal, single or multiple; Paleoc.-Recent.

Subfamilies: 3

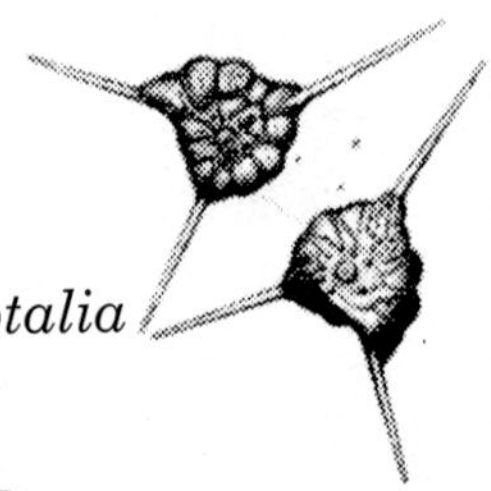

Asterorotalia

Elphidinae: Sutural pore and canal system present, with retral processes bridging sutures; aperture interiomarginal or areal and multiple; Paleoc.-Rec. e.g. *Elphidium*

Elphidium

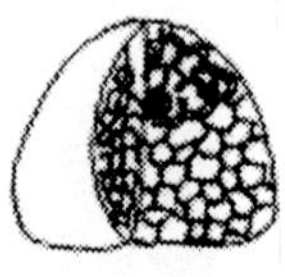

Miogypsina

Lepidorbitoidae: Bilocular embryonic chamber followed by equatorial chambers that may be spatulate, hexagonal or arcuate and differentiated lateral chambers; no canal system; U. Cre.-L. Eoc.

Subfamilies: 3

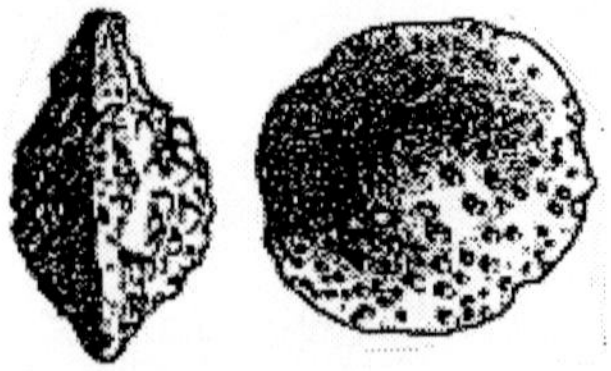

Miogypsina

Miogypsinidae: Test with appressed laminae well developed lateral chambers at both the side of equatorial layer of chambers; embryonic chambers followed by epiembryonal chambers; spiral and interseptal canals present; stolons connect equatorial chambers; Oligo.-Mio. e.g., *Miogypsina*

Superfamily | **Family Characters and Distribution**

Nummulitacea:

Planispiral involute to evolute, lenticular, discoidal to globular; chambers numerous, subdivided into chamberlets, may be come annular; septal flaps like Rotaliacea; spiral marginal cord and spiral canal system present, may be modified or replaced by intraseptal canals in advanced forms; U.Cre.-R.

Discocylina

Operculina

Families: 4

Nummulitidae: Test planispiral, numerous median chambers; simple or subdivided into chamberlets; complex canal system of septal, marginal and vertical canals; aperture arched, slitlike at the base of apertural face; U.Cre.-Rec. e.g., *Nummulites*.

Subfamilies: 3

Pseudoorbitoididae: U. Cre. Subfamilies: 3

Discicyclinidae: Initial chamber of embryonic apparatus embraced by larger second chamber in megalospheric forms; later chambers planispiral which enlarge to become annular and subdivided into chamberlets, annular and radial stolons present; interseptal and intramural canal system present; L. Eoc.-M. Eoc. e.g., *Discocyclina*

Orbitoclypeidae: Like Discocyclinidae but with several layers of small lateral chambers, annular chambers not secondarily subdivided into chamberlets; Paleoc.-Eoc.

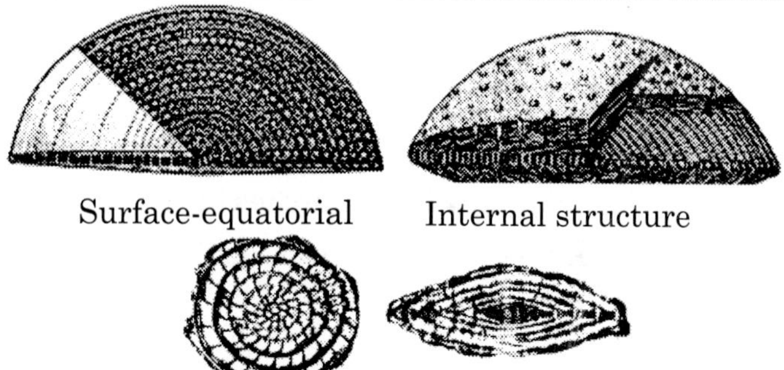

Surface-equatorial Internal structure

Equatorial Section *Axial Section*

Nummulites (also see Fig. 3.20)

Suborder CARTERININA

Trochospiral , of secreted spicules, each of which is a single elongated crystal of calcite, held by poorly cemented organic matrix; which disintegrates, therefore, taxa have meager fossils record; Eoc.-Rec.

Superfamily	**Family characters and distribution**
None	Family: 1
Carterina	**Carterinidae**: Same as the suborder; Eoc.-Rec. e.g. *Carterina*.

3.6 ECOLOGY

Ecology is a study of reciprocal relation or interaction between individuals and their physical and chemical surroundings. It reveals the causes of patterns of distribution and abundance of organisms together with the interaction between individuals and their physical and chemical environments, interactions between species and also interaction between individuals of the same species (Murray, 1991). The ecological studies are carried out through field-studies, laboratory experiments or mathematical modelling.

The paleoecology of the organisms, however, is derived out of the ecological parameters of their living counterparts. As it is based on an assumption that 'the present is the key to the past'; it is pertinent to all of us to answer two questions: (i) how representative are the modern environments? and (ii) are the ecological relationships seen in modern faunas applicable to fossil record? In case of extinct-organisms such interpretations are derived through the ecological aspects of other living species, the fossil counterparts of which are found in association with the taxa.

Murray (1991) found that in the last few centuries man caused environmental changes by way of farming and industrialization that released loose sediments, fertilizers, sewage, nutrients, chemical-pollutants. Similarly, issues like 'post glacial sea-level rise' in the past 10,000 yrs and 'origin of oceanic bottom waters' are controlled by shapes and depths of ocean basins, which are results of plate tectonic and climatic processes, may be addressed to.

The principles behind their paleoecologic interpretations, and age dating involves a combination of basic concepts of sedimentology, ecology and oceanography (Setty, 1979), coupled with abundant distributional data for living species that have been generated through wide and intensive researches in the last five decades.

Foraminifera occur in nearly all the marine and paralic environments. Their abundance may be judged by a fact that about 1000 to 2,500,000 foraminifera live/sq m in the modern marine basins.

Most of them are benthic, living upon the sea floor, within the upper few centimeter of ooze or upon the benthic algae or other organisms. Spectrum of their presence in marine environments may be visualized by their occurrence in varying depths (intertidal zone to deep oceanic levels); varying salinities (brackish to hypersaline) and varying latitudes (tropic to the poles).

Walton's (1952) 'Rose of Bengal' method is used to differentiate protoplasm bearing tests (recently living) from empty (dead) tests. It has revolutionised the study of ecology of benthic species. Ecological studies of foraminifera constitute a dominating feature of oceanographic investments. They are preferred as compared to the other microfossils owing to their interpretative taxonomy and proven utility in the studies related to environment and climate of the present and past, petroleum exploration, biostratigraphy, high resolution biostratigraphy, biostratigraphic correlations, tracing up of the stratigraphic boundaries, paleolatitudinal reconstruction, marine transgression and regressions, study of oceanographic aspects like chemical, physical and biological parameters of environments, sediments transfer, up welling, major oceanic currents, etc.

Although a vast literature exists on the modern benthic species there is still much scope left to work out the basic distribution patterns. Recently many culture studies, mathematical modelling, stable isotope composition of the wall, and ultrastructure, biochemistry of soft parts and genomics have thrown new light on various aspects of their physiology, mode of life and mode of preservation.

The niche

Niche is often thought in spatial or habitat sense and commonly from the data collected at a single point in time (Lee, 1974) as shown in Fig. 3.14. Very little is known about the realized niches of any group of foraminifera from any locality of the world.

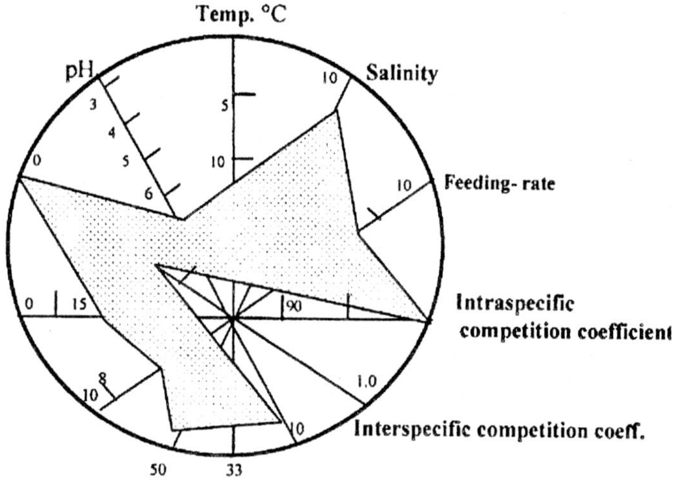

Figure 3.14 Graphic representation of realised-niche of *Spiroloculina hyalinea* (Lee, 1974, *in* Murray, 1991)

The factors that control distribution of organisms may be primarily grouped into two: aboitic (temperature, salinity, dissolved oxygen availability, pH) or biotic (food, intra- and interspecific competition, predation, etc.). They may also be categorized as the physical, chemical and biological environmental factors (Kureshy, 1968) and are collectively or individually responsible for the distribution of benthic foraminiferal species. The benthic species give more clues for the environment, depth, temperature, bottom-conditions and salinity while the planktic species, about the temperature, current turbulence and turbidity and distribution of stable isotopes of the surface waters.

The 'physical environmental factors' include temperature and the depth of water; light penetration; geomorphology of the ocean floor; water movements; sediments (substrate); sediments transfer; geographical position and shape of water body; oceanic currents; turbidity; and turbulence.

The 'chemical environmental factors' include salinity; organic content of the sediments; and the gases found dissolved viz. pH (hydrogen ion concentration); carbon dioxide, oxygen; hydrogen sulphide contents; and the Eh (oxidation-reduction potential).

The 'biological factors' include the associated organisms; symbiotic relation with other species; and food supply.

Although it is quite obvious that the environmental factors, as mentioned above do not operate independently yet for the sake of simplicity their influence on the distribution of foraminifera have been discussed individually :-

Physical environmental factors

Temperature of water

It is one of the most important factors for all the marine organisms particularly the protozoan like foraminifera. Vaughan (1940) classified the temperature zones of the modern oceans in to five categories, given below:

Latitudinal Zone	Temperature Range
Polar zone	-1.9° C to +5°C
Sub-polar zone	5° C to 10°C
Temperate zone	Up to 25°C
Subtropical zone	15°C to 33°C
Tropical zone	25°C to 36°C

The culture studies have clearly established that each species is adapted to a specific range of temperature for its survival, growth and reproduction. Also that in warm water condition the metabolism is rapid, and alternation of generation remains normal. Therefore, under such conditions the evolution is expected to attain a maximum rate.

If other factors remain normal the agglutinated forms dominate the cold water conditions whereas the calcareous forms do not attain their normal growth and size. Likewise, the large calcareous forms, like rotaliids and miliolids dominate the warm water environments while the agglutinated forms remain under developed.

Temperature controls the ecology of foraminifera due to its horizontal (geographica/latitudinal zonation) and vertical distribution (depth/bathymetric zonation). Thus, the role of temperature appears to affect both the distribution and the morphology of the organisms:

(a) Influence on distribution: The surface layer of 200m undergoes a seasonal variation but in the range of 200-100 m or more, that is the permanent thermocline the temperature decreases rapidly with the depth.

Therefore, there may be a kind of stratification of progressively cooler layers of water up to 4°C in the abyssal plains while the surface temperature remains nearly 28°C. Eventually, the cooler layers are characterized by benthic forms dominated by the agglutinated tests and underdeveloped calcareous forms which are otherwise found in polar regions.

McGlasson (1959) found that the bathymetric zonation of living forms, along the southern California coast has 3 depth zones:

Depth Zone	**Temperature Range**
0-37m	11.1°-20.6°C
38-74m	9.6°-14.2°C
75-180m	8.0°-11.1°C

The temperature can not be solely responsible for the distribution of genera as there are a number of forms e.g., *Elphidium,* which live in shallow water at all latitudes. Krasheninnikov (1960) stated that different faunal species occur at the same depth, depending on the rate of sedimentation and bottom characters. Boltovskoy and Wright (1976) opined that the cause of vertical or bathymetric zonation of foraminifera may not be solely attributed to the temperature distribution as it is a result of combination of several factors that are related to depth of water.

(b) Influence on test morphology: Temperature plays an important role in the development of variation in a few forms (Carpenter, 1856). Culture-experiments show that variation in the size, number of chambers, size and spacing of pores, presence and absence of secondary pores in planktic forms are some of the morphological changes in same species, caused by temperature variation.

The forms adapted to the cooler denser water bodies have fewer buoyancy problems and have lower test-porosity than those from the warmer or shallower waters (Frerich, *et al.* 1972). The latter improve their buoyancy with high test porosity and prominent spines. In an interesting study of Pleistocene sequence fluctuations in temperature has been found on the basis of dominance of dextral (clock-wise or right handed direction of coiling) in *Globigerina pachyderma* and *Globorotalia truncatulinoides* in warmer waters and *vice versa* (Funnell and Riedel, 1971).

Said (1950) found that the depth zonation of foraminifera of the Red Sea is controlled by organic content and substrate, but not by temperature.

In summary, it seems as though (1) temperature controls the vital activities of foraminifera. (2) each species has certain tolerance limits as well as a temperature optimum. (3) a species can only tolerate temperature extremes for a short period. (4) it is the primary factor which controls the geographical as well as vertical distribution of species. (4) it influences the test size and morphology.

Depth of water

It is a very important ecological factor but may influence foraminifera because of depth related changes that occur in other environmental parameters such as temperature, pressure etc. Increase in depth is inversely proportional to temperature, light intensity, oxygen and directly proportional to pressure, and carbonate solubility. Due to the increase in depth the benthic assemblage of littoral, neritic, bathyal and abyssal environment vary considerably. Such effects of variation in depth are more pronounced in calcareous forms in comparison to the agglutinated ones. Depth also causes some changes in the morphology of forms for e.g., the species may become more rounded, more ornamented or shorter in size with every considerable increment in depth.

Norton (1930) gave the depth zonation of foraminifera of Florida and West Indies region (Table 3.2).

Table 3.3 Depth zonation off Florida and West Indies (based on Norton, 1930)

	Depth (m)	Temp. (°C)	Distribution of Families
Zone A	0-10	20 – 30	Dominance of Miliolidae & Peneroplidae. Valvulinidae, Calcarinidae, Homotrematidae & Nonionidae are frequent.
Zone B	10-110	20 – 25	Decrease in Miliolidae & Peneroplidae. Nodosariidae, Textulariidae, Buliminidae & Asterigerinidae are frequent.
Zone C	1100-1500	4.0 - 7.5	Dominance of Globigerinidae & Globorotaliidae. Nodosariidae & Buliminidae thrive, but are less abundant.
Zone D	3500-5000	1.8 - 2.0	Dominance of Globigerinidae & Globorotaliidae. Nodosariidae & Buliminidae thrive, but less abundant

Benthic foraminiferal faunas are zoned offshore according to depth of water (Phleger 1960). He proposed the faunal depth boundaries on the continental shelf to slopes and basins but, Cooper (1964), Funnell (1967) and others showed that the depth is not the only controlling factor.

A progressive increase in depth causes lowering of temperature and increase in the solubility of carbon dioxide. Further, the ratio of CO_2 to O_2 also increases with depth because algae can not photo-synthesis below the photic-zone, although the animals continue to respire. This leads to decrease in pH with depth from about 8.2 to as low as 7.0. The level at which the $CaCO_3$ solution equals $CaCO_3$ supply is called the 'calcium carbonate compensation depth' or CCCD. It lies between 4,000-5,000 m in the Pacific ocean but varies from latitude to latitude. In polar-regions it occurs at 400-500 m. The species living below the CCCD show effects of dissolution on tests, less in benthics than planktics. The dominance of agglutinated species increases below the CCCD.

Saidova (1967) discussed the details of the distribution of benthic and planktic foraminiferal assemblages in the successive depth zones of different latitudes:

(a) Benthic foraminiferal zones:

i. Miliolidae and Nummulitidae dominate up to 200m depth; between 20°N to 20°S latitudes; in the tropical and some part of southern subtropical regions.

ii. *Amphistegina* dominates up to 500 m depth, between 40°N to 40°S latitudes, (tropical and some part of southern subtropical region); and up to 1500 m, between 20°N to 20°S (northern subtropical and the remaining part of southern subtropical region).

iii. *Trochammina* and *Cyclammina* dominate up to 1500 m and 2500 m, respectively, between 60°S to 80°S (Antarctic region); up to 2000 m between 20°N to 60°S (tropical, southern subtropical and cold temperate region); and up to 2000 m (in the cold temperate and northern subtropical region).

v. *Alabamina* and *Uvigerina* occur up to 4500 m between 20°N to 60°S (tropical, southern subtropical and cold temperate region) and up to 3500 m or more (in cold temperate and Antarctic region).

vi. *Rhabdammina*, *Bathysiphon*, *Cyclammina*, *Homosina* and *Haplophragmoindes* may occur up to 6000 m depth, especially in the tropical, southern subtropical and cold temperate regions.

(b) Planktic foraminiferal zones:

As mentioned earlier also the tests of *Globigerina pachyderma* coil sinistrally in the cold temperate (0° to 15°C) and Antarctic (0° to 5°C) regions; and dextrally in the northern subtropical (18° to 24°C) and cold temperate (5° to 18°C) regions.

The proportion of planktic to benthic species in modern-sea helps determining water-depth gradients in ancient rocks, in a rather general way. The modern planktic species increase sharply offshore, reaching a maximum on continental slope and deep water basins. Benthic, however, dominate in continental-shelf sediments and near shore regions; calcareous on the selves, and arenaceous species in near shore brackish-environments like marshes and lagoons (Fig. 3.15). The ratio of planktic vs. calcareous and arenaceous benthonic species changes with increasing water depth from near shore environment across the continental shelf and down the continental slope (Fig. 3.15).

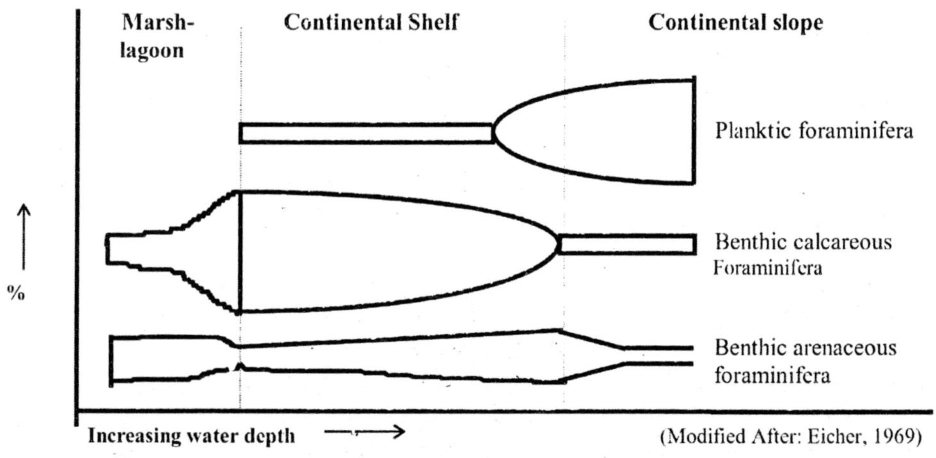

Figure 3.15 Changing proportions of planktic, calcareous benthic and arenaceous benthic species with increasing depth

Light

It is useful to all photosynthesizing organisms in 'photic-zone' but is checked by several factors such as turbidity, turbulence, depth of water etc. Since the growth in algae and phytoplanktons support the foraminiferal populations by symbiotic relation or by providing direct nutrition, the growth of foraminifera is also adversely affected if the light intensity is checked in the 'near-shore' or 'highly-turbid' areas.

Current

Due to minute size, their tests may be sorted, reworked and carried away by currents in the same way as a mineral grain of same size and dimension. Therefore, the coarse sediments have higher population percentage than those which are entirely composed of silt and clay (Kureshy, 1958).

Turbulence

The wave-action in the littoral-zone, especially along the beaches, and up to of 100 m depth causes strong turbulence in the water body, and does not allow the settlement of smaller particles like mud, clay and silt. Such conditions may restrict the occurrences of the foraminifera as their pores are choked.

Turbidity

In the high turbidity zones the penetration of light is severely affected as the particulate matter remains suspended in water for a longer time, and up to a greater depth. It also checks the process of photosynthesis in micro-algae and phytoplanktons which is otherwise allowed up to 120 m depth in clear tropical water; 40 m in shelf waters; and 15 m or so in coastal waters of temperate latitudes. It is seen that the calcareous populations are severely affected, and dominated by larger and stronger species like *Amphistegina.*

Substrate or bottom sediments

The substrate plays a very important role in distribution and morphology of benthic foraminifera. A direct proportional relationship was noted between foraminiferal number and the median diameter of the sediment, in which they live (Said, 1950). Therefore coarse sediment support more benthic species while silt and clay support less number of species Kureshy (1958). Optimum sites for foraminifera are sands and fine sands mixed with a considerable quantity of mud (Saidova, 1961). Kathal (1989) noticed dominance of spinose benthic forms like *Asterorotalia trispinosa* in the clayey and muddy substrates, along the east coast of India.

Substrate also influences the test wall of the arenaceous forms. For example *Textularia conica* of fine-grained sediments is agglutinated by finer grains than those which thrive on the coarser sediments.

Substrate may range from firm surface, like rocks, shells, and seaweeds or other plants, to soft unconsolidated sediments and ecologically three levels may be recognised (Murray, 1991):

(i) Epifaunal: species living on or above the sediment surface as sessile or attached, clinging or attached mobile and free;

(ii) Semi-epifaunal: living partly below the sediment surface as sessile, clinging or free

(iii) Infaunal: living within the sediments as sessile, clinging or free.

There seems to be some relationship between mode of life and test architecture. Species without septa and one aperture, hemispherical, zigzag (*Tolypammina*), planispiral (*Ammodiscus*, *Cornuspira* and *Sprillina*) and branched forms (*Dedrophyra*) in septate species low-trochospiral with broad umbilical area (*Cibicides*, *Rosalina* and *Discorbis*) belong to epifaunal mode but low trochospiral, biconvex (*Amphistegina*), large species, Family Soriditae to clinging epifaunal mode (Brasier, 1980).

Table 3.4 Selected examples of mode of life in benthic foraminifera (after Murray, 1991)

Habitat:	**Epifaunal**	**Semi-epifaunal**	**Infaunal**
Sessile	*Cibicides, Rosalina* etc.		
Clinging	*Rosalina, Planorbulina, Glabretella, Patellina, Cancris, Hanzawia, Pararotalia, Elphidium* etc	*Saccammina, Marsipella Bathysiphon*	*Trochammina, Elphidium*
Free	*Elphidium, Peneroplis, Quinqueloculina, Triloculina*		*Elphidium, Ammonia, Bolivina, Bulimina, Reophax, Textularia, Uvigerina*

Currents and tides

Several effects of currents and tides may be summarized as below:

(a) The bottom distribution of both living and dead tests is physically altered;

(b) The benthic forms can be transported into the planktic zones;

(c) The number of living specimens can be adversely affected due to rapid movements (tidal zone); and

(d) The normal test morphology may be altered.

Chemical environmental factors

Salinity

Salinity plays a very important role in the ecology of marine organisms particularly the protozoan like foraminifera owing to their unicellular nature and very high susceptibility to the minute environmental changes. The majority of them occur in normal saline water, a very small percentage is found in brackish water, and a very few are found in fresh water environment. Hilterman (1949) classified the water masses on the basis of salinity as:

Category	**Salinity** (‰)
Fresh water	0-5
Miohaline	3-5
Mesohaline	5-9
Pliohaline	9-16.5
Brachyhaline	16.5-30
Sea water	30-40

A few additional terms like mixohaline (18-30‰); hypo- (< 30‰); hyper- (30-40‰); and ultrasaline (> 40‰) are also in practice.

Like temperature, 'salinity tolerance limits' for survival, growth and reproduction of species can be determined. Such limits in benthics are relatively wider and distinct in comparison to the planktics. Normal range of salinity (30-40‰) seems to have little effect on the former as most of them survive even in mixohaline waters (18-30‰).

The considerable mixing of fresh and saline water in coastal zones makes the populations quantitatively and qualitatively different from those areas characterized by normal marine salinities. As regards the planktics, conditions are uniform as they usually occur within 34-36‰ range, but can tolerate it up to 40‰. Some species can tolerate wide limits of salinities e.g., marsh-species and may occur in both, brackish and hypersaline waters. With lowered salinity their number is reduced but stenohalines increase greatly. Extreme variation in salinities affect the reproductive cycle of the species resulting in the retardation of growth, size, thickness and ornamentation of the test. Considerable variation in salinity of water up to 200 m and slight variations even up to 100-1500 m depth are noticeable. However, the salinity becomes nearly constant below 1500 m. The low salinity of brackish lagoons and marshes favours the 'low diversity assemblage' of agglutinated fauna (most with non-labyrinthic walls and organic, siliceous or ferruginous cements e.g., *Reophax* and a few Rotaliacea e.g., *Ammonia*). The soft tectinous e.g., *Allogromia* lives in fresh and brackish water, but rarely gets preserved. The high $CaCO_3$ concentration of hypersaline waters favour porcelaneous (Miliolina, especially Nubeculariidae and Miliolidae). Murray (1973) has proposed a triangular model for the interpretations of the salinities or paleo-salinities. The relative proportions of Textulariina, Miliolina and Rotaliina are plotted to show the tentative salinities conditions (Fig. 3.16).

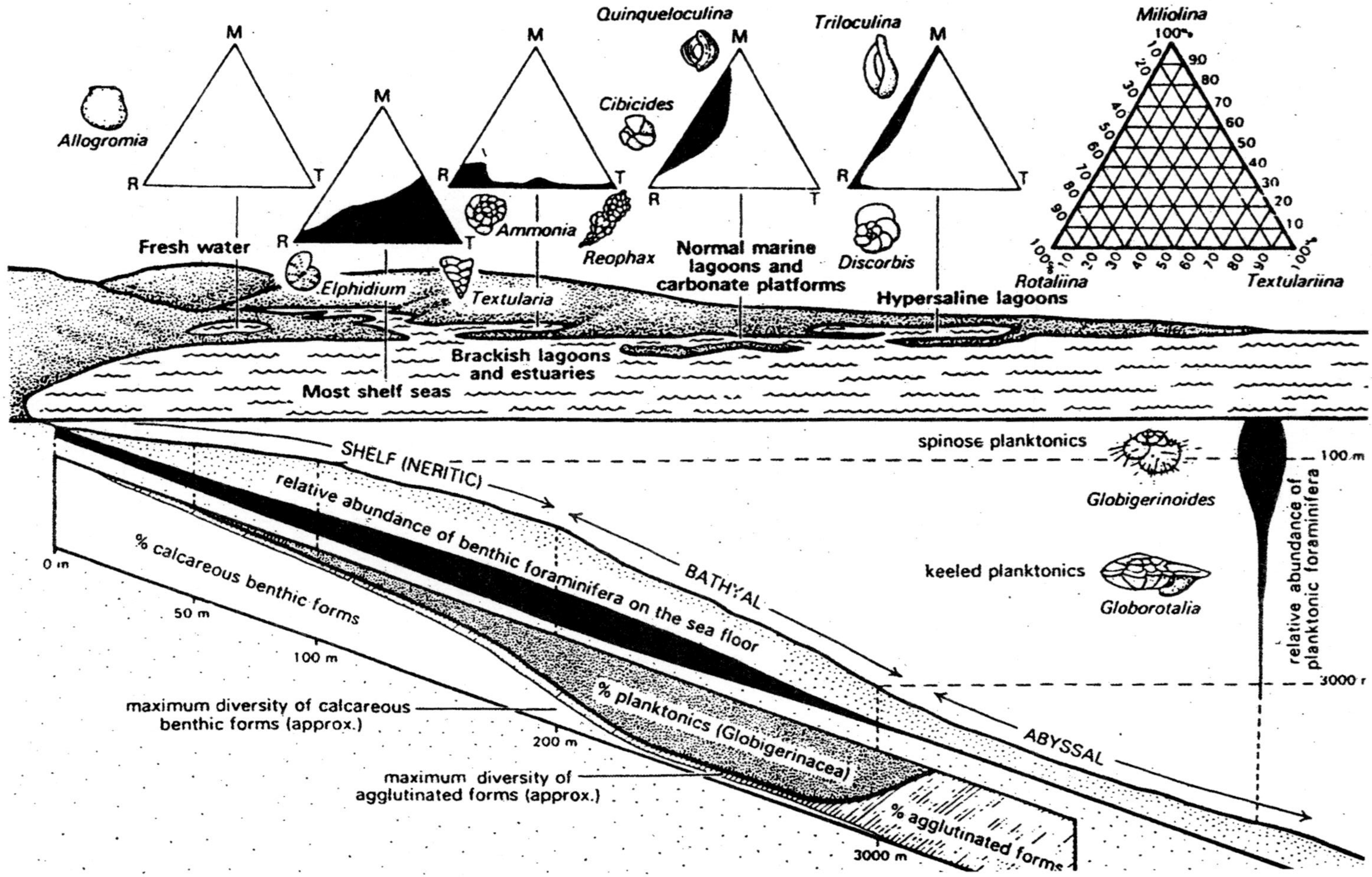

Figure 3.16 Abundance of benthic and planktic Foraminifera: change in composition of assemblages due to change in salinity (Murray, 1973); and due to change in depth and salinity (Brasier, 1980).

It may be summarized as though the salinity is directly related to their vital activity. Each species has certain critical and as well as optimum salinity-limits for various physiographic functions. It also affects their geographical distribution. Because of the relatively uniform distribution of salinity in most of the open seas, this effect is much more pronounced in areas where there is considerable mixing with fresh water. The seasonal variation in salinities also affects their distribution and occurrence. It can also affect test morphology, primary size and ornamentation of tests. Generally reduction in size and loss of ornamentation is observed near the salinity tolerance limits.

Organic content of the substrate

Foraminifera feed on small bacteria, algae, protists and the other organic matter and an adequate supply of these nutrients is of the utmost importance to them. The taxa, which maintain symbiotic relations with algae, are benefited to a great extent as compared to the non-symbiotic ones. It is believed that the larger foraminifera could have attained their size in this manner.

A seasonal bloom in nutritive conditions may result in bloom of foraminifera and *vice versa*. The planktics are influenced by the presence and seasonal bloom in the population of phyto-planktons in the photic zone, whereas the benthic assemblages by the presence and concentration of organic content of the substrate.

In general, higher content of organic matter supports the growth of populations, but if it brings the pH of water down due to decay, it may retard the same. The effects of pH lower than 7 (acidic water) has been dealt separately. Phleger (1960) has cited numerous cases along the lagoons and bays in which the higher organic content has enhanced the growth even in finer substrates.

Dissolved gases

The concentration of oxygen in water (4.0-6.5 ml/l) allows respiration in organisms and oxidation of the organic matter. The photic-zone remains rich in oxygen content due to the photosynthetic process. A bloom in population of phytoplanktons and rise in temperature (during summer months) enhance the oxygen content of water.

The poor oxygen content affects the foraminiferal population in two ways: the size or shape of test is reduced or modified; may help in abundant development of Bolivinidae. Some benthic species e.g., *Rotorboides granulosum* has lived in the oxygen-depleted, shallow and protected environments ever since its appearance in Middle Pliocene, Late Miocene ? (Kathal, 1989 and Kathal and Bhalla 1996 a and b).Presence of hydrogen sulphide (H_2S), produced through the bacterial decay of organisms at the bottom sediments make the environment reducing, and inhospitable for most of the foraminifera. H_2S gets further concentrated due to absence of oxygen at such depths.

Carbon dioxide and hydrogen ion concentration (pH)

The carbon dioxide enters in marine waters either by its dissolution from atmosphere or organism's respiration and exists in the system through photosynthesis by the autotrophic protophytes, algae and plants. The excessive CO_2 lowers pH of the water, which is otherwise maintained within 7.8 - 8.3. The pH tolerance limits of different species are different:

At pH 7.8, dissolution of calcareous tests starts. As a result, the organisms have to spend considerable energy in recalcifying tests especially in the marshes and bays where such conditions form due to decay of organic substance in presence of oxygen.

Calcium carbonate ($CaCO_3$)

The concentration of $CaCO_3$ in sea water depends upon the temperature, salinity and pressure of the water. The temperature increases the salinity, hydrostatic pressure and solubility of the bicarbonate ions. The $CaCO_3$ content, therefore, plays a vital role in the development of larger tests in tropical climate. In general, it influences the ecology of foraminifera in two ways: (a) It controls the distribution of foraminifera as a consequence of its availability; and (b) It influences the size and robustness of calcareous tests.

In cold climate/deeper zones of tropical climate, however, where the secretion of $CaCO_3$ is not fully allowed the population is either dominated by 'thin-walled' calcareous tests or agglutinated species. With a progressive increase in depth, the temperature of water decreases but the ratio of CO_2 to O_2 increases as the photosynthesis is not allowed below the photic zone, although the animals continue to respire. This in turn leads to decrease in pH, from about 8.1 to as low as 7.0 vis-à-vis dissolution of $CaCO_3$, at the deeper levels. As discussed earlier the level at which the dissolution of $CaCO_3$ equals its supply is called the 'calcium carbonate compensation depth' or CCCD. It varies from latitude to latitude for e.g., it is found between 4000-5000 m in Pacific but at 400-500 m in Polar Regions. The species living below CCCD show effects of dissolution on the tests. Such effects are less in benthic forms in comparison to the planktic ones. The populations below the CCCD are dominated by agglutinated species.

Biological environmental factors

As discussed earlier the foraminifera are by and large omnivorous. However, several 'phytal' species live closely associated with marine algae where the carbon dioxide is utilised by the algae and the released oxygen is consumed by foraminifera. The relation helps them attaining large size.

Some 'endosymbiont species' allow microscopic algae to live below the inner wall of the test, showing a marked 'symbiotic relation', between the two. Hallock (1999) found that out of 150 living familie, less than 10% harbour algal endosymbionts. Four miliolid, three rotaliid and five globigerinid families host algal endosymboionts (Hallock, 1999). Besides, several other families sequester chloroplast, which is harvested from algal food, for days to weeks after ingestion.

Ornamented Peneroplidae hosts symbionts from red algae; unornamented Peneroplidae and species of Soritidae hosts green algae; Soritinae hosts dinoflagellates and planktic foraminifera host both red-green and green as symbiont. Although this is not true symbiosis, yet this phenomenon does enable foraminifera to benefit directly from photosynthesis. Three rotaliid families Amphisteginidae, Calcarinidae and Nummulitidae host diatom symbionts.

3.7 GEOLOGICAL APPLICATION OF FORAMINIFERAL ECOLOGY

Paleoenvironment is deduced using the ecological parameters, like (a) the living counter parts of fossils contained by them; and (b) the 'associated-species', in case of extinct fossils. Such studies are carried out in two ways:

(i) The study of ecology of the individual species, called autecology; and

(ii) The study of ecology of communities, called synecology.

There are several factors in their ecology that are used to study the paleoenvironment of marine and brackish waters, especially of the Tertiary stratas that are the promising horizons for petroleum exploration. The researches of the last 60 years have resulted in selecting some of the ecological factors as tool in oil-exploration mainly for identification of the source rocks for hydrocarbons, in sedimentary basins.The evidences of paleoecology are divided into 'physical', 'chemical' and biological categories.

The salient features of application of their ecology are discussed below:

Morphological variation

(a) The marshy fauna is dominated by simple agglutinated taxa. The beach fauna, where the environment is characterized by turbulence, temperature variation, evaporation, run-in and run-off of precipitated water and unstable substrate is often characterized by broken, coarsely perforate, calcareous forms. The lagoonal facies of rocks is characterized by similar euryhaline fauna and dominated by species like *Ammonia, Elphidium* and *Quinqueloculina*.

(b) The brackish water fauna generally consists of cosmopolitan species. 'Outer bay facies' and 'inner bay facies' may be demarcated on the basis of ratios of '*Ammonia-Elphidium*' and '*Ammonia-Ammobaculites*' (or calcareous-agglutinated). Ecophenotypes of calcareous species help delineating 'outer-bay' and 'inner-bay' facies in much better way in comparison to the 'calcareous-agglutinated' ratio, required in petroleum exploration (Kathal, 1996).

(c) The coastal shelf fauna is dominated by benthic species. The planktic forms first out-number them in the 'outer-shelf area', and then completely replace them in the upper bathyal zone. Therefore, a Planktonic/Benthic (P/B) ratio of nearly 20:1 is an indicative of a depth not less than 200 m. The distribution patterns of benthic, planktic, calcareous, and agglutinated species indicate the marsh, beach, channel, lagoon, inner and outer neritic, bathyal and abyssal regions, respectively.

(d) The simple agglutinated species indicate bay and lagoonal environments. Their labyrinthic walled species, with siphonate-chambers occur in the central- and outer-shelf, and the bathyal environments. The diverse miliolids and the biloculine dominate in the inner-shelf and the bathyal regions, respectively. The smooth-walled, calcareous-perforate species occur in the inner shelf; the simple-ornamented species (costate and striate) in the outer shelf and the bathyal environments; and the coarsely ornamented, larger and thicker forms in the deeper zones.

(e) In the abyssal and hadal-zones the CCCD prevents the accumulation of calcareous tests considerably. As such these regions are fully represented by highly diversed agglutinated species.

(f) Kathal (2002a and 2000b) noted relation of change in the morphology of the Recent benthic foraminifera on the basis of environmental parameters, along the enrire length of the east coast of India. The comparative morphological of the Indian Recent and Japanese species shows that the two otherwise distant regions have a moderate faunal affinity (Kathal and Singh, 2011; Singh and Kathal, 2011).

(g) Cosmopolitan planktic species (*Globorotalia menardii*) show remarkable change in the frequency as it occurs as 'rare' species in the latitudes north of 18°N. To the south it becomes 'common', especially in the regions of 'south equatorial currents', and 'abundant' in the equatorial waters. They are good 'paleothermometers' of the past climate e.g., *Globigerina bulloides* dominates in the slope-waters during summers and, gets replaced by *Globigerinoides ruber* in winters. The fauna of 'lower core' in tropical regions may be similar to modern fauna of higher latitudes.

(h) The phytoplankton productivity controls the standing crop of planktic foraminifera. The other factors like salinity, temperature, oxygen content, phosphates etc., are secondary. The majority of the species living at the surface or within the euphotic zone are spinose e.g., *Globigerina*, while those living/lived in the lower zones, are/were commonly thicker or keeled e.g., *Globotruncana*.

Ecological applications

The ecological parameters of the benthic foraminifera have been correlated at micro-habitat levels. Corliss (1985) noted the importance of benthic foraminiferal ecology of the species within deep-sea sediments. Also that distribution certain species is not directly controlled by overlaying bottom-water conditions but within the sediments. Some other case studies are as follows:

(a) The ratio of living and dead fauna in the Recent sediments indicates the rate of sedimentation in the area. Therefore, a decrease in the relative number of living specimens points to a decrease in the rate of sedimentation.

(b) Nigam and Khare (1999) identified two morpho-groups, 'angular-asymmetrical' and 'rounded-symmetrical'. The former were less in number in a river mouth due to high river discharge and low-salinity of water and but the latter were dominating.

(c) Based on coherent variation in mean-proloculus diameter and abundance, it was postulated that it can also be used to infer increased seasonality in organic matter production, showing variation in strength duration of monsoon (Saraswat *et al.*, 2005).

(d) Increased number and diversity of foraminifera (than the last recorded data) in the Zuary estuary showed an improvement in the environmental health of the estuary (Panchang *et al.*, 2005).

(e) Nigam *et al.* (2005) found a drastic fall in total foraminiferal number and diversity in the lower reaches of the Mandovi River estuary due to suspended-load released from to mining activities in the catchment area of the Mandovi River estuary.

(f) Salinity response of benthic foraminifera may be used for environmental assessment as *Pararotalia nipponica* shows that comparatively lower-salinities are much more detrimental to the foraminiferal tests than higher salinities (Nigam *et al.*, 2006).

(g) Kathal (2005) highlighted the probability of enviroenmental impact of the controversial upcoming 'Sethusamudram Ship Canal', on the marine life of the Gulf of Mannar and Palk Bay regions, south-eastern Coast of India, based on formaminiferal ecology.

(h) A study, based on their ecology showed that that in addition to the global Holocene sea level rise, tectonic vertical displacement is the cause of the destruction of the soft coral-patches, off the west coast of Myanmar (Panchang *et al.*, 2006).

(i) Q-mode cluster analysis, off Kachchh and radiocarbon dating revealed an inverted sequence (fine-grained sediments and small foraminifera sandwiched between normal) postulating that ~8,000 years ago sediments ranging in age from ~10,000 to ~12,000 years B.P. were eroded from deeper offshore deposits by storm/tsunami (Nigam *et al.*, 2006).

(j) A study showed persistence of benthics under low dissolved oxygen conditions predates anthropogenic influence (Nigam *et al.*, 2009).

(k) A culture-study of *Rosalina leei* showed that hyper-salinity that causes its dissolution is recoverable under normal salinity up to a normal adult-size. (Sujata *et al.*, 2011).

(l) *Rosalina globularis* showed its tolerance to temperature-variations (20°- 35°C) and variations in its morphology under different climatic conditions (Saraswat *et al.*, 2011)

(m) The 'coiling direction ratio' in some species is a significant stratigraphic tool. The dominance of right handed (dextral) or left handed (sinistral) forms may change suddenly in a sedimentary sequence, thus giving a clue of non-discontinuous sedimentary environment. The ratio has been succes-sfully used recently in deciphering the paleoclimate by Nigam and Khare (1992).

(n) Saraswati *et al.* (1993) studied the Palaeogene isotopic temperatures of western India, based on foraminiferal fauna.

(o) Ghosh *et al* (2008) studied the implications intertidal foraminifera in the macrotidal estuaries of the Gulf of Cambay for interpreting sea level change in palaeo-estuaries.

(p) Saraswati (2007) gave an important account of the symbiont-bearing benthic foraminifera of Lakshadweep.

Paleomonsoonal studies

In Asian the monsoonal rainfall is of prime importance for economy of the countries. The foraminiferal ecology may be used in the interpretation of paleomonsoonal precipitation.

Some important studies carried on the paleomonsoonal studies, based on foraminiferal data are discussed below:

(a) An increase in the Asian monsoon was recorded during the past four centuries based on study of *Globigerina bulloides* (Anderson, Overpack and Gupta, 2002).

(b) Nigam and Khare (2004) recorded signals of the change in monsoonal precipitation at around 2,000 year BP in a sediment core, off the west coast of India.

(c) Gupta *et al.* (2008) found a relatively well oxygenated Oxygen Minimum Zone (OMZ) in the early Holocene characterised by the influence of intense monsoon-related production that was mitigated by increased circumpolar deep water (CDW) incursion based on $\delta^{18}O$ isotope of *Uvigerina peregerina*, the microhabitat of which shifted across the mid Holocene transition.

(d) A study of planktonic species showed that the Mid-Pliocene climate change was intensified by swift in Indonesian subsurface flow (Gupta *et al.*, 2009).

(e) Ghosh and Saraswati (2008) noted palaeomonsoonal upwelling on the western coast of India based on the studies of *Gallitellia*.

Stratigraphic correlation

Correlation of foraminiferal fauna of two different regions, in time and space is useful in biostratigraphy. The two important basis of correlation of planktic fauna are: (a) their worldwide contemporaneous existence, and (b) their unchanging sequential relation to one another.

Paleontological affinities of the smaller foraminifera of the eastern and western hemisphere even during the Eocene times have been noted. In Oligocene time also, ties between fauna of the Atlanto-Caribbean and eastern Pacific regions and in Miocene in between Australia, New Zealand and Fiji has been noted. Gibson (1989) pointed out that the tubulogeriniids (benthic foraminifera) migrated from the Atlanto-Caribbean to the Eastern Pacific ocean in late Oligocene-Miocene times. Kathal and Bhalla (1996a and b) found that *Rotorboides granulosum* and other associated species of Recent foraminifera of Panamanian province originated and very well established in Atlanto-Caribbean region prior to the Early Pliocene (Miocene?).

Table 3.5. Species characterising marginal marine environments (after Murray, 1991)

Table 3.5 A. Shelf environments

Agglutinated Test	Porcelaneous Test	Hyaline Test	
1. *Clavula cylindrical*	1. *Cornuloculina balkwilli*	1. *Acervulina inhaerens*	15. *Elphidium subarcticum*
2. *Cribrostomoides jeffreysii*	2. *Cyclorbiculina compressa*	2. *Ammonia falsobeccarii*	16. *Epistominella vitrea*
3. *Eggerella advena*	3. *Massilina secans*	3. *Bolivina pacifica*	17. *Gavelinopsis praegeri*
4. *Eggrelloides medius*	4. *Miliolinella subrotunda*	4. *Bolivina seminuda*	18. *Glabratella ornatissima*
5. *Eggrelloides scaber*		5. *Bolivina spissa*	19. *Loxostomum pseudobeyrichi*
6. *Lepidodeuterammina ochracea*		6. *Brizalina striatula*	20. *Nonionella stella*
7. *Saccammina atlantica*		7. *Bulimina aculeata*	21. *Nonionoides scaphum*
8. *Textularia earlandi*		8. *Bulimina marginata*	22. *Planorbulina mediterranensis*
9. *Textularia sagitulla*		9. *Bulimina tenuata*	23. *Rosalina anomala*
10. *Textularia truncate/T. bocki*		10. *Cassidulina laevigata*	24. *Rosalina globularis*
		11. *Cassidulina reniforme*	25. *Spirillina vivipara*
		12. *Cassidulina teretis*	26. *Suggrunda eckisi*
		13. *Chilostomella ovoidea*	27. *Uvigerina curticosta*
		14. *Cibicides lobatulus*	28. *Valvulineria bradyana*

Table 3.5 B. Carbonate environments (larger foraminifera, after Murray, 1991)

Porcelaneous Test	**Hyaline Test** (all have diatom symbionts)	
1. *Alveolinella quoyi*	1. *Amphistegina bicirculata*	11. *Calcarina hispida*
2. *Amphisorus hemprichii*	2. *Amphistegina gibbosa*	12. *Calcarina spengleri*
3. *Archaias angulatus*	3. *Amphistegina lessonii*	13. *Cycloclypeus carpenter*
4. *Borelis pulchra and schlumbergeri*	4. *Amphistegina lobifera*	14. *Heterocyclina tuberculata*
5. *Cyclorbiculina compressa*	5. *Amphistegina papillosa*	15. *Heterostegina operculinoides*
6. *Marginopora vertebralis*	6. *Amphistegina radiata*	16. *Neorotalia calcar*
7. *Parasorites orbitolinoides*	7. *Baculogypsina sphaerulata*	17. *Operculina ammonoides*
8. *Peneroplis planatus*	8. *Baculogypsinoides spinosus*	18. *Operculina complanata*
9. *Sorites marginalis*	9. *Calcarina defrancii*	19. *Operculina cummingii*
10. *Sorites orbiculus*	10. *Calcarina gaudichaudii*	20. *Operculinella venosus*

Table 3.5 C. Deep-sea species (after Murray, 1991)

Agglutinated Test	Hyaline Test	
1. *Adercotryma glomeratum*	1. *Brizalina albatrossi*	18. *Globobulimina affinis*
2. *Astrorhiza arenaria*	2. *Brizalina ordinaria*	19. *Globobulimina auriculata*
3. *Bathysiphon capillare*	3. *Bulimina inflata*	20. *Globobulimina pyrula*
4. *Bathysiphon filliformis*	4. *Bulimina marginata*	21. *Hoeglundina elegans*
5. *Cribrostomoides subglobosus*	5. *Cancris auriculus*	22. *Melonis barleeanum*
6. *Cyclammina cancellata*	6. *Cassidulina carinata/laevigata*	23. *Melonis zaandamae*
7. *Cystammina pauciloculata*	7. *Cassidulina neocarinata*	24. *Nonionella iridea*
8. *Hormosinella guttifer*	8. *Chilostomella oolina*	25. *Nuttallides pusillus*
9. *Reophax bilocularis*	9. *Chilostomella ovoidea*	26. *Nuttallides umboniferus*
10. *Reophax scorpiurus*	10. *Cibicides refulgens*	27. *Oridorsalis umbonatus*
11. *Rhabdammina abyssorum*	11. *Cibicidoides kullenbergi*	28. *Rupertia stabilis*
12. *Saccorhiza ramose*	12. *Cibicidoides robertsonianus*	29. *Trifarina angulosa*
	13. *Epistominella exigua*	30. *Trifarina bradyi*
	14. *Epistominella rugosa*	31. *Uvigerina elongatastriata*
	15. *Fontbotia wuellerstorfi*	32. *Uvigerina mediterranea*
	16. *Fursenkoina Mexicana*	33. *Uvigerina peregrine*
	17. *Gevelinoposis translucens*	

Table 3.6. Ecology of different benthic genera based main living occurrences (after Murray, 1991)

Order of data in the table: mode of life; substrate; feeding; salinity; temperature; depth; environment. Definitions: hard substrates= rocks, shelf macrofauna, plants, etc.; S= salinity, brackish 0-32, marine 32-37, hypersaline >37; temperature of the bottom water, cold, temperate, warm (even in tropical areas the water deeper than 100 m is temperate are cold); environment, shelf 0-180 m, bathyal 180-4000 m, abyssal >4000 m.

Table 3.6A. Genus of mainly epifaunal and/other habitats

Genus and their ecological details
1. ***Acervulina***: epifaunal, attached; braozoa, shells, etc.; passive suspension feeder; temperate to warm; 0-60m; inner shelf.
2. ***Adercotryma***: epifaunal or infaunal, free; fine sand; detritivore; marine; cold; shelf-abyssal.
3. ***Alabaminella***: epifaunal, free; often associated with phyto-detritus but also lives on sediment; normal marine; cold; deep sea.
4. ***Alveolinella***: epifaunal, clinging; algal-covered carbonate gravels; herbivore, symbionts; 18-26°C; 5-100 m; inner shelf, lagoon.
5. ***Alveolophragmium***: epifaunal, free; sand; detritivore; 10°C; 20-700 m; shelf-upper bathyal.
6. ***Amphisorus***: epifaunal, clinging; phytal, carbonate sediment; herbivore, symbionts; 19-26°C; 0-50 m; lagoon, near-shore.
7. ***Amphistegina***: epifaunal, free; phytal, coarse carbonate; herbivore symbionts; winter minimum 15°C; generally >20°C; 0-130 m; coral reefs, lagoons.
8. ***Archaias***: epifaunal, clinging; phytal; herbivore, symbionts; >22°C; 0-20 m; inner shelf.
9. ***Arenoparrella***: epifaunal and infaunal, free or clinging; muddy sediment; herbivore; brackish-hypersaline; 0-30°C; mainly intertidal on marshes but some on shelf-upper bathyal.
10. ***Articulina***: epifaunal, free; phytal or sediment; marine-hypersaline; >20°C; 0-3500 m; inner shelf-bathyal.
11. ***Asterigerina***: epifaunal, free; sediment; herbivore?; 0-100 m; subtropical-tropical; inner shelf.
12. ***Asterigerinata***: epifaunal, free; sediment; herbivore?; temperate; 0-100 m; inner shelf.
13. ***Asterorotalia***: epifaunal, free; sediment; herbivore?; subtropical-tropical; inner shelf.
14. ***Astrorhiza***: epifaunal, sessile; sediment; passive suspension feeder/carnivore; cold-temperate; shelf.
15. ***Baculogypsina***: epifaunal, clinging; phytal; herbivore, symbionts; >25°C; 0-10 m; coral reefs, lagoons.
16. ***Balticammina***: epifaunal, free; detritivore?; very euryhaline; uppermost tidal marine.
17. ***Bigenerina***: epifaunal?, free; sediment; omnivore; temperate-warm; shelf.
18. ***Bolivina***: infaunal-epifaunal, free; muddy sediment; some tolerate dyse detritivore?; cold-warm; inner shelf- bathyal.
19. ***Borelis***: epifaunal, free; algal-coated substrates; sea-grass, coarse sediment; herbivore, symbionts; 18-26°C; 0-40 m; lagoon-reef.
20. ***Calcarina***: epifaunal, free; sediment; herbivore, symbionts; 18-26°C; 0-30m; lagoons, reefs, inner shelf.

Contd... 98p.

Table 3.6 A. Genus of mainly epifaunal and/other habitats (*Contd.*)

Genus and their ecological details
21. ***Cancris***: epifaunal, free; sediment; detritivore?; marine; temperate-subtropical; 50-150 m; shelf.
22. ***Astrononion***: epifaunal-infaunal, free-clinging; muddy sediment; detritivore?; marine; cold; shelf-bathyal.
23. ***Cibicides***: epifaunal, attached; hard substrates in high energy; passive suspension feeder?; marine; cold-warm; 0-2000 m; shelf-bathyal.
24. ***Cibicidoides***: infaunal epifaunal, clinging; hard substrates; passive suspension feeder?; cold; shelf-bathyal.
25. ***Cuneata***: infaunal, free; mud, sand; detritivore; S 30-35; cold; shelf.
26. ***Cribrostomoides***: epifaunal, free or clingling; sand, pebbles; detritivore; S 30-35; <15°C; shelf-bathyal.
27. ***Ehrenbergina***: epifaunal?, free; mud; detritivore?; marine; cold; outer shelf-bathyal.
28. ***Elphidium***: (Keeled species) epifaunal, free; sand, vegetation; herbivore; temperate-warm; S 30-70; 0-50 m; inner shelf; (non-keeled species) infaunal, free; mud; sand; herbivore, S 0-70; brackish-hypersaline marshes and lagoons, inner shelf (upper bathyal, *Elphidium excavatum* only).
29. ***Epistominella***: epifaunal to shallow infaunal, free; mud, phytodetritus; detritivore; temperate-cold, shelf-bathyal.
30. ***Eponides***: epifaunal, free or clinging; sediment or hard substrates; detritivore?; cold-temperate; shelf-abyssal.
31. ***Fontbotia***: epifaunal, clinging; hard substrates; passive suspension feeder in currents; cold; bathyal.
32. ***Gaudryina***: epifaunal, attached; hard substrates; suspension feeder?; temperate; 50-500 m; shelf-upper bathyal.
33. ***Gavelinopsis***: epifaunal, clinging; hard substrates; passive suspension feeder?; cold-temperate; shelf-bathyal.
34. ***Glabratella***: epifaunal, clinging or attached; hard substrates; herbivore?; marine-hypersaline; temperate-warm; 0-50 m; hypersaline marshes and lagoons, marine inner shelf.
35. ***Goesella***: epifaunal; free; sediments; detritivore; temperate-warm; shelf-upper bathyal.
36. ***Gyroidina***: epifaunal, free; mud; detritivore?; cold; shelf-bathyal.
37. ***Hanzawaia***: epifaunal, clinging; hard substrates; passive suspension feeder?; temperate-warm; inner shelf.
38. ***Cyclammina***: epifaunal?; free; mud, sand; detritivore; >100 m; outer shelf-abyssal.
39. ***Cycloclypeus***: epifaunal, free; muddy; carbonate sediment; warm; 60-100 m; shelf, below storm-wave base.
40. ***Deuterammina***: epifaunal, attached mobile; sediment; normal marine; cold-temperate; shelf.

Contd... 99p.

Table 3.6 A. Species of mainly epifaunal and/other habitats (*Contd.*)

Genus and their ecological details
41. ***Discorbis***: epifaunal, clinging or attached, hard substrates, coarse sand; herbivore; temperate-warm; 0-50 m; inner shelf.
42. ***Haplophragmoides***: epifaunal to shallow infaunal, free; mud-sand; detritivore?; cold-temperate; marshes-bathyal.
43. ***Hemisphaerammina***: epifaunal, attached; hard substrates; passive herbivore; cold-temperate; inner shelf.
44. ***Heterolepa***: epifaunal; clingling?; hard substrates; passive suspension feeder; cold-temperate; shelf-bathyal.
45. ***Heterostegina***: epifaunal, free; phytal muddy carbonate sediment and hard substrates; herbivore, symbionts; 18-26°C; 0-130 m; shelf, lagoon, seeks shade.
46. ***Homotrema***: epifaunal attached; hard layers, high energy; passive suspension feeder; temperate-warm; inner shelf, reefs
47. ***Hyalinea***: epifaunal, free; mud; silt; detritivore?; marine; cold-temperate; 10-1000m; shelf-upper bathyal.
48. ***Jadammina***: epifaunal, free; mud, silt; herbivore? Or detritivore?; S 0-50; 0-30°C; high intertidal marsh.
49. ***Karreriella***: epifaunal, free; mud, silt; detritivore?; marine; <10°C; outer shelf-bathyal.
50. ***Lepidodeuterammina***: epifaunal-infaunal, clinging; hard substrates, coarse sand, high energy; herbivore; cold; shelf.
51. ‘***Trochammina***’ : epifaunal/infaunal, free; sediment; herbivore/detritivore; S 0-60; 0-30°C; 0-6000 m; intertidal-abyssal.
52. ***Trochulina***: epifaunal or infaunal, free; sediment; detritivore?; temperate-warm; lagoons; shelf.
53. ***Uvigerina***: mainly infaunal, some epifaunal, free; muddy sediments; detritivore?; cold; 100-4500 m; shelf-abyssal.
54. ***Lenticulina***: epifaunal, free; mud; detritivore?; marine; cold; outer shelf-bathyal.
55. ***Marginopora***: epiphytic, attached; phytal, Carbonate sand and gravel; herbivore, symbionts; 18-26°C; 0-45 m; inner shelf, lagoon, coral reefs.
56. ***Martinottiella***: epifaunal?, free; mud; detritivore?; cold; >120 m; outer shelf-bathyal.
57. ***Massilina***: epiphytic, clinging; plants; herbivore; temperate-warm; inner shelf.
58. ***Miliammina***: infaunal-epifaunal, free; mud, silt, decaying vegetation; detritivore; brackish-hypersaline; 0-30°C; marshes-upper bathyal.
59. ***Miliolinella***: epifaunal, clinging; plants and hard substrates; herbivore; S 32-50; 10-30°C; 0-100 m; hypersaline lagoons, normal marine lagoons and marshes, inner shelf; some deep-sea records.
60. ***Neoconorbina***: epifaunal, clinging; hard substrates; herbivore?; marine; temperate; inner shelf.

Contd... 100p.

Table 3.6 A. Species of mainly epifaunal and/other habitats (*Contd.*)

Genus and their ecological details
61. ***Nuttallides***: epifaunal-shallow infaunal, free or clinging, detritivore?; <4°C; bathyal-abyssal.
62. ***Operculina***: epifaunal, free; carbonate sediment; herbivore, symbionts; slightly hypersaline; warm; 0-130 m; lagoon-shelf, low energy and medium light.
63. ***Operculinella***: epifaunal, free; carbonate sediment; herbivore, symbionts; warm; 0-130 m; lagoon to shelf.
64. ***Oridorsalis***: epifaunal, free; mud; detritivore; <4°C; bathyal.
65. ***Pararotalia***: epifaunal, free; sand; herbivore; warm; inner shelf.
66. ***Patellina***: epifaunal, clinging; hard substrates; herbivore?; cold-warm; 0-100 m; inner shelf.
67. ***Peneroplis***: epifaunal, clinging; plants and hard substrates: S 35-53; 18-27°C; 0-70 m; lagoons and innermost shelf.
68. ***Planorbulina***: epifaunal, attached; hard layers, high energy; suspension feeder?; temperate-warm; 0-50 m; inner-shelf.
69. ***Planulina***: epifaunal, clingling hard; substrates; passive suspension feeder?; marine; cold-warm; shelf-bathyal.
70. ***Poroeponides***: epifaunal, clinging; hard substrates; herbivore/passive suspension feeder?; temperate-warm; inner shelf.
71. ***Portatrochammina***: epifaunal, free; sediment; detritivore; marine; cold-temperate; shelf-bethyal.
72. ***Pullenia***: infaunal, free; mud; detritivore; marine; cold; outer shelf-bathyal.
73. ***Pyrgo***: epifaunal, free/lining; plants/sediments; herbivore; temperate-warm; inner shelf ; or epifaunal, free; sediment, detritivore?; cold; shelf-bathyal.
74. ***Quinqueloculina***: epifaunal, free or clinging; plants or sediments; herbivore; marine-hypersaline; S 32-65; cold-warm; hypersaline lagoons; marine marsh and shelf, rarely bathyal.
75. ***Recurvoides***: epifaunal, free; sediment; detritivore; marine; cold; shelf-bathyal.
76. ***Rosalina***:epifaunal, clinging/attached; hard- layers; herbivore?/omnivore; temp.-warm; up to 0-100m; lagoons, inner mid-shelves.
77. ***Saccammina***: infaunal, free; sand; detritivore; marine; cold-temperate; 0-100m; inner-mid shelf.
78. ***Siphotrochammina***: epifaunal, sometimes epiphytic on mangrove roots; brackish; warm; tropical, marsh and mangrove.
79. ***Sorites***: epiphytic, clingling; plants, mainly sea-grass; herbivore, symbionts; S 37-45; 18-26°C; 0-70 m; lagoons, near-shore.
80. ***Spirilina***: epifaunal, clinging; hard substrates; marine; cold-temperate; 0-100m; inner-mid shelf.
81. ***Spirolina***: epiphytic, clingling; plants; herbivore; S 37-50; 18-28°C; lagoons, near-shore.
82. ***Spiroloculina***: epifaunal, free/clinging; sediment/plants; herbivore; marine-hypersaline; temperate-warm; 0-40 m; lagoons, inner shelf.
83. ***Textularia***: epifaunal, free or clingling; hard substrates, sediment; detritivore?; cold-warm; 0-500 m; lagoons, shelf-bathyal.
84. ***Tiphotrocha***: epifaunal, free; sediment; herbivore or detritivore; brackish; temperate; intertidal; marsh.
85. ***Triloculina***: epifaunal, free or clinging; mud, sand, plants; herbivore, detritivore; marine-hypersaline; S 32-55?; temperate-warm; mainly hypersaline lagoons or marine inner shelf, some bathyal species (cold).

Table 3.6 B. Species of mainly infaunal and/other habitats

Genus and their ecological details
1. ***Ammoastuta***: infaunal?, free; sediment; detritivore?; brackish; temperate-tropical; marshes.
2. ***Ammobaculites***: infaunal, free; muddy sediment; detritivore; brackish-marine; temperate-tropical; brackish marshes and lagoons; inner shelf-upper bathyal.
3. ***Ammoglobigerina***: infaunal?, free; sediment; detritivore?; marine; cold; shelf-upper bathyal.
4. ***Ammonia***: infaunal, free; muddy sand; herbivore?; brackish, marine, hypersaline; warm temperate-tropical; 0-50 m; brackish and hypersaline lagoons, inner shelf.
5. ***Ammoscalaria***: infaunal?, free; sediment; detritivore; brackish-marine; cold-temperate; inner-shelf.
6. ***Ammotium***: infaunal, free, muddy sediment; detritivore; brackish; 0-30°C; 0-10 m; tidal marshes, brackish lagoons & estuaries, enclosed brackish shelf
7. ***Aubignyna***: infaunal?, free; sediment; slightly brackish; temperate to warm; lagoon inner shelf.
8. ***Bathysiphon***: infaunal, free?; sediment; marine; cold; upper bathyal.
9. ***Bolivinella***: infaunal?, free; muddy sediment; some species tolerate-warm; shelf.
10. ***Bolivinellina***: infaunal, free; muddy sediment; marine-slightly brackish; temperate inner shelf and marginal marine.
11. ***Brizalina*** : infaunal, free; muddy sediment; some species tolerate dysoxia; detritivore?; marine; cold-temperate; marginal marine-bathyal.
12. ***Buccella***: infaunal?, free; muddy sediment; detritivore?; marine cold-temperate; 0-100 m; lagoons-inner shelf.
13. ***Bulimina***: infaunal, free; mud-fine sand; few species tolerate dysoxia; detritivore: cold-temperate; inner shelf-bathyal.
14. ***Buliminella***: infaunal, free; muddy sediment; detritivore?; marine; temperate; upper-shelf but also lagoons and upper bathyal.
15. ***Cassidella***: infaunal, free; mud; detritivore; temperate; shelf.
16. **Cassidulina**: infaunal or epifaunal, free; mud, sand; detritivore; marine; cold-temperate; shelf-bathyal.
17. ***Chilostomella***: deep infaunal, free; mud; tolerance dysoxia; detritivore; upper-shelf bathyal.
18. ***Discorinopsis***: infaunal, free; muddy sediment; detritivore; hypersaline; warm; marsh.
19. ***Eggerella***: infaunal, free; silt, fine sand, detritivore; S 20-37; cold-temperate; shelf-outer lagoon.

Contd...102p.

Table 3.6 B. Species of mainly infaunal and/other habitats (*Contd.*)

Genus and their ecological details
20. ***Eggerelloides*** : infaunal, free; sand; detritivore; temperate; shelf.
21. ***Eilohedra***: infaunal, free; mud; detritivore?; cold; bathyal.
22. ***Elphidiella***: infaunal?, free; sand; detritivore?; S 30-35; cold; shelf.
23. ***Fursenkoina***: infaunal, free; mud; some species tolerate dysoxia; detritivore; S 30-35; shelf- upper bathyal.
24. ***Globocassidulina***: infaunal, free; mud; detritivore?; cold-temperate, shelf-bathyal.
25. ***Globobulimina***: deep infaunal, free; facultative anaerobe; mud; tolerates dysoxia; detritivore; cold-temperate; shelf-bethyal.
26. ***Haynesina***: infaunal, free; mud, silt, herbivore; brackish, cold-warm; marsh, lagoon, inner shelf.
27. ***Hoeglundina***: infaunal, free; mud; detritivore?; cold; outer shelf-bathyal.
28. ***Hormosina***: infaunal, free; mud; detritivore?; cold; bathyal-abyssal.
29. ***Islandiella***: infaunal, free; mud; silt; detritivore?; <10°C; >20 m; shelf-upper bathyal.
30. ***Logenammina***: infaunal?, free; sediment; detritivore?; S 30-35; cold-temperate; shelf-bathyal.
31. ***Melonis***: infaunal, free; mud, silt; some species tolerate dysoxia; detritivore; cold; shelf-bathyal.
32. ***Nonian*** : infaunal, free; mud, silt; herbivore; S 30-35; cold; 0-180 m; shelf.
33. ***Nonionella***: infaunal, free; mud; some species tolerate dysoxia; detritivore?; temperate-warm; 10-1000 m; shelf-upper bathyal.
34. ***Nonionellina***: infaunal, free; mud-sand; detritivore; cold; shelf.
35. ***Nonionoides***: infaunal, free; muddy sediments; some species tolerate dysoxia; detritivore?; temperate-warm; shelf.
36. ***Reophax***: infaunal, free; mud, sand; detritivore; marine; cold-temperate; lagoons; shelf-bathyal.
37. ***Trifarina***: infaunal, free; mud, sand; detritivore?; marine; cold-temperate; 0-400 m; shelf-upper bathyal.

Paleontological zonation

The foraminiferal studies coupled with the radioactive dating have helped in establishing the concept of high resolution biostratigraphic zonation and correlation. The correlation derived by comparing the faunal zonation with faunal components between nearby sections is reasonably accurate, especially when planktics are involved. Berggren (1972) proposed a refined radiometric time-scale for the Cenozoic to which the planktic zonation can be related.

Boundary problems

The Cretaceous-Tertiary boundary has witnessed: (a) the major extinction phase of animal and plants; (b) the changes in flora and fauna; and (c) the appearances and disappearances of several foraminiferal species. The patterns of water circulation in the Cenozoic oceans have occurred due to the opening of the North Atlantic along the Raykjames Ridge. Their increased diversity can be correlated to the rise in temperature during this period.

The observations at the boundary between the Mesozoic/Cenozoic (or Cretaceous/ Tertiary or Masstrichtian/ Danian, or Masstrichtian/Lower Palaeozoic) reveals that the Danian was less than 1 m thick and the entire Paleocene was only about 5 m thick in South Atlantic marked by an abrupt change in fauna and flora. Likewise, the boundary problems of Paleocene-Eocene, Miocene-Pliocene have been solved.

The correlation of climatic data based on the planktic foraminifera has helped in solving the problem of Pliocene-Pleistocene boundary, on the basis of ratio of coiling directions of *Globigerina pachyderma, Globorotalia truncatulinoides, Pulleniatina obliquiloculata* and *Globoquadrina conglomerata. Hyalinea balthica*, a cold water indicator, is yet another important marker of this boundary owing to its appearance in the beginning of Pleistocene.

Kathal and Singh (*in editorial hands*) substantiated the assignment of Danian age to planktonic foraminifera in the Intertrappean near Chhindwara, Central India and solving the K/T Boundary attempted earlier by Keller *et al.* (2009) and Khosla *et al* (2009).

Paleotemperature and isotope stratigraphy

Oxygen isotopes

In the evaporating phase, O^{16} migrates to the surface with slightly a greater velocity. Thus the water gets enriched in heavier O^{17} and O^{18}. Since the ocean water is being evaporated continuously, it ought to have greater proportions of O^{17} and O^{18}. Urey (1947) showed that the relationship between O^{18} and O^{16} (O17 negligible) in calcareous shells depends on the temperature (paleotemperature) of water in which the calcium carbonate is secreted.

Emiliani (1955) determined the range of fluctuations in oceanic temperature through the analysis of calcareous foraminiferal tests in a sedimentary sequence.

The empirical relationship (Berger and Gardner, 1975) between the temperature of precipitation and isotopic composition is:

$$T = k_1 - k_2 - (\delta - A) + k_3 (\delta - A)^2$$

where

$$\delta = \frac{O^{18}/O^{16}\,(\text{sample}) - O^{18}/O^{16}\,(\text{Standard carbonate})}{O^{18}/O^{16}\,(\text{standard carbonate})}\,10^3$$

and

$$A = \frac{O^{18}/O^{16}\,(\text{water}) - O^{18}/O^{16}\,(\text{Standard water})}{O^{18}/O^{16}\,(\text{standard water})}$$

where k_1 =16.5; k_2=4.3; and k_3 = 0.14 (Savin and Douglas, 1973)

The oxygen isotope equilibrium denotes similar environmental conditions; hence, a difference in this ratio in different species means different temperatures during their growth (Savin and Douglas, 1973).

The oxygen isotope studies shows: i. The temperature declined very rapidly at the end of the Cretaceous; ii. The bottom water temperature in South Atlantic was 10°C (higher in the upper Cretaceous than the Recent); and iii. The Antarctic ice-sheet was already present at its present dimensions in the Pliocene.

Fig. 3.17 shows the ratio of O^{16} and O^{18} in the ocean water in glacial and interglacial periods.

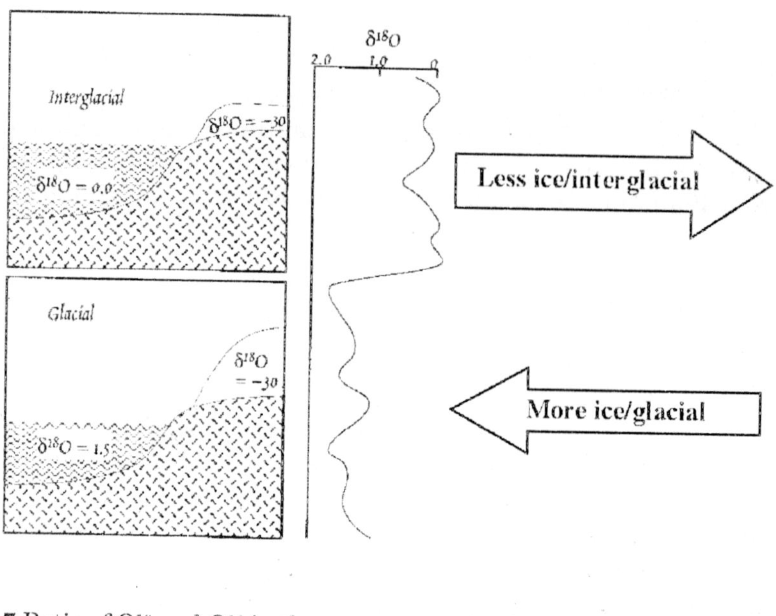

Figure 3.17 Ratio of O^{16} and O^{18} in the ocean-water in glacial and interglacial-periods

Carbon isotopes

The behaviours of carbon isotopes are less well understood. However, the lighter isotope of carbon (δC^{12}) is preferentially removed from the sea water by planktic organisms. After death, foraminiferal organic matter decays in deep water as a result of which these waters become progressively rich in this isotope. Therefore, the measurement of carbon isotope record in benthic forms provides history of deep water circulation changes and large scale variations in the cycling of carbon between atmosphere, biosphere and ocean. Srinivasan (1996) documented important interpretations of isotopic studies of oxygen and carbon conducted by Kennett and Stott (1991); and Keller (1985):

a. Abrupt changes in oxygen and carbon isotopic composition of planktic and benthic foraminifera in the latest Paleocene (56 to 55 Ma) and Mass extinction in deep sea benthic foraminifera.

b. Surface temperature during early Eocene was 19°C. It declined to 7°C in Oligocene, though separated by episodes of temperature stability.

c. Fluctuations in the Pleistocene temperature.

d. The early Neogene climatic optimum marked by the lowest values of O^{18} isotope.

e. Lowering of oceanic temperature during middle Miocene and cryspheric evolution in Antarctica.

f. Establishment of depth stratification of Miocene planktic foraminifera, development of thermal gradients and evolution of modern biostratigraphic patterns in planktic species.

Strontium isotopes

Hodell *et al.* (1987) in Srinivasan (1996) pointed out that variation in Sr-isotopic ratios hold significant promises for addressing several long standing problems:

i. Refining of the late Neogene biostratigraphic datums.

ii. Upgraded correlation and dating of the late Neogene epoch boundaries

iii. Correlation of near-shore and deep sea sequences and dating of transgressive and regressive cycles on continental margins.

iv. Correlation of palaeooceanographic events.

3.8 APPLICATIONS IN HYDROCARBON EXPLORATION

The role of foraminifera in petroleum exploration may be fully appreciated by understanding a few aspects of the 'reservoir rocks' and 'source rocks' and their depositional environments; basin evolution etc., using:

i. Correlation and environmental analysis techniques (detailed correlations, unconformities and sediment accumulation);

ii. Higher resolution biostratigraphy, sequence- and seismic stratigraphy;

iii. Physical and chemical environmental analysis, and the factors controlling their ecology; and

iv. Problems related to biostratigraphic analysis.

The paleontology primary aims at the formulation of biostratigraphic-zones of Phanerozoic sequences on the basis of their fossil contents. So far as mega fossils are concerned, they may be found only at outcrops and not in the bore-hole samples that are required in the present day exploration techniques for hydrocarbons/other minerals. Whereas, 'microfossils' are numerous in such subsurface samples, e.g., more than 20,000 fossils of foraminifera, thousands of Ostracoda and millions of nanno planktons and others microfossils may be obtained from per kilogram of sediments, rock-surfaces, borehole-cores and cuttings.

Foraminifera for their long geological history, wide geographic distribution and exceptional diversity of test composition, highly visible external morphology, form, and structure and rapid evolution could be studied for various applications. Their unique susceptibility to minute environmental changes makes them useful in the studies on environment and ecology of the present and past. Oceanographically, they play a widespread role as they are exclusively marine (except a few) and their distinct assemblages thrive in all the possible marine, and marginal marine environments irrespective of geographical and bathymetric boundaries in time and space. Their more than 35,000 living and dead species lend themselves to statistical analysis. Many benthic forms have more than one morphological state per species. Murray (1991) listed characteristic species, belonging to the different oceanic micro-environments.

The primary environments required for conversion of organic matter into oil are:

(i) Sediments if deposited at high-rate (generally in estuaries), cover organisms before they are bacterially destroyed/oxidized and thus favour formation of oil;

(ii) Usually a high-current-index is not favourable for generation of petroleum as it allows flow of oxygen that oxidizes the oil-forming organisms;

(iii) High-degree of reducing conditions prevents oxidation of organisms and favours conversion of organic-matter into oil; and

(iv) Compaction of sediments during palingenesis is favourable for the formation of an oil-pool.

The first commercial application of microfossils especially foraminifera began in late nineteenth century when the middle Miocene age was assigned to well-hole samples in Vienna basin, Austria. J.A. Udden introduced their importance in developing high-resolution biostratigraphy and biostratigraphic correlation.

In the early days, foraminiferal- biostratigraphy sometimes did not match with the chronological age of strata until Blow, a British petroleum paleontologist in 1969, proposed the standard 'Tertiary Planktic Foraminiferal Zonation' using lineages of selected planktic species. The progress in their use in oil-exploration gained momentum in the last 4 decades by in development of:

(i) High resolution, and multi-microfossil biostratigraphy;

(ii) Delineation of stratigraphic boundaries;

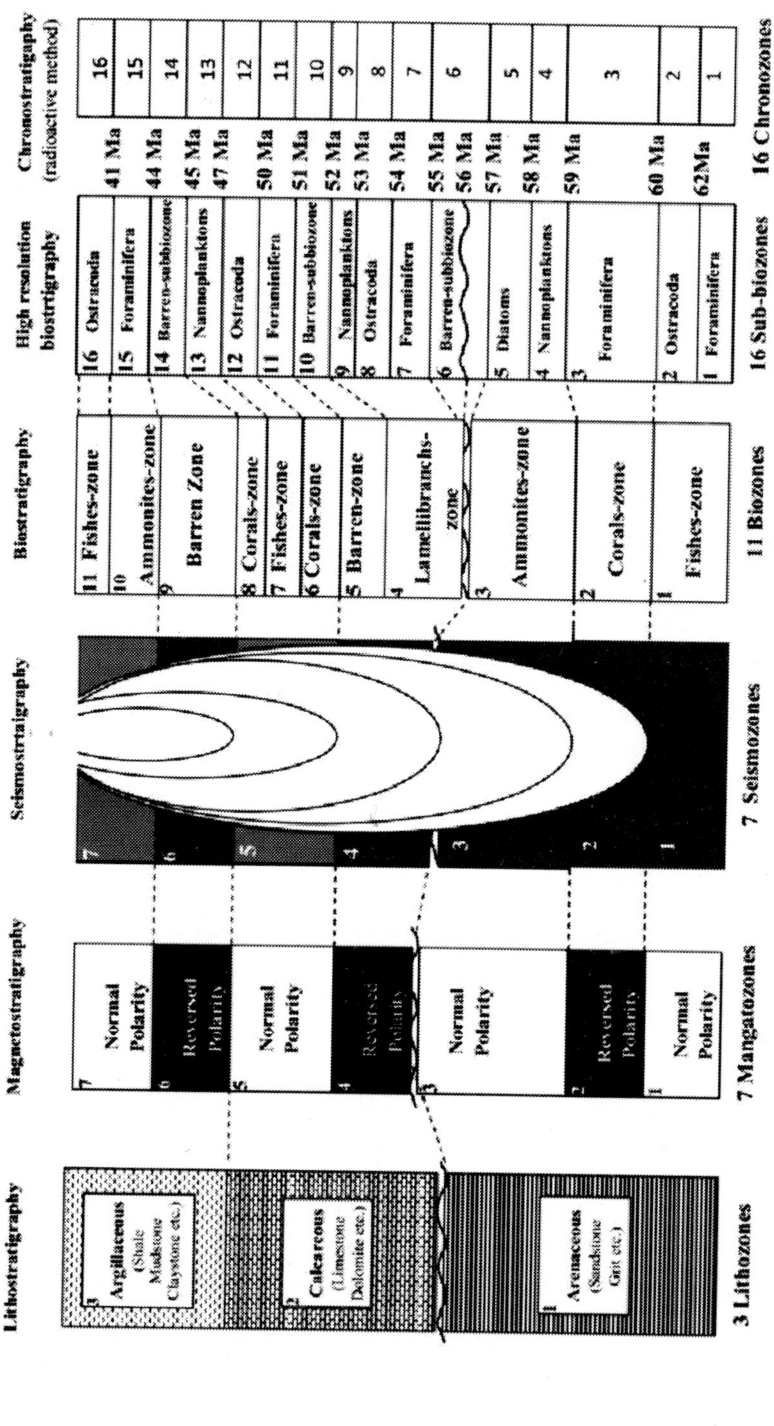

Figure 3.18 Possible stratigraphic-zones in an imaginary succession based on possible zoning-criteria with imaginary chronological ages

(iii) Development of sequence stratigraphy;

(iv) Interpreting sea-level changes and paleodepths;

(v) Reconstruction of 'redox' environment;

(vi) Paleoslope analysis;

(vii) Development of integrated synergic biostratigraphic and geochemical approaches in petroleum systems; and

(viii) Preparation of 'subsurface images'.

This phase of studies gave rise to many new petroliferous basins attributable mainly to use of scanning electron microscope (SEM) of old species together with the enumerable species encountered in different deep-sea projects (DSDP, IPOD and now ODP).

Their zonations for oil-exploration are based on: a) characteristics 'planktic assemblages' to study 'deep marine facies'; b) 'larger benthics' to date 'shallow marine sections'; c) and 'smaller benthics' to reconstruct 'finer paleodepth zones'; and d) 'sea level changes' or 'paleoenvironmental modelling'. The benthics also help 'well to well' correlation in regions with 'stratigraphic boundary problems'.

The advances in well logs and seismic stratigraphy have apparently eroded the importance of micropaleontology to some extent; however, they cannot replace it.

The major challenges in oil-exploration are:

(i) Slicing, dating and correlation of the stratigraphic column is possible only in selected deep marine facies where planktic foraminifera are common. In shallow marine sections larger species like *Globotruncana* and *Miogypsina* are used for dating;

(ii) Making of the First Appearance Datum (FAD's) of species has limitation especially in cutting samples.

(iii) Finer paleodepth zonation of marine shelves, slopes or basin floors is to be fully achieved through uvigeriniids and other smaller benthic foraminifera to deduce 'sea level changes' and reconstruct 'paleoenvironmental models'.

Some concepts of foraminiferal use in oil-exploration

As discussed (Chapter 3) the benthic species live on or within top few mm of marine sediments (substrate) in various micro-environments and may provide much information about the physical, chemical and biological parameters like water depth, salinity, temperature, current-characters etc., Planktic species help studying physical, chemical and biological aspects of the surface waters and the depositional environments and global biostratigraphic correlations. Some important aspects of their parameters are:

1. Biostratigraphy based geologic history of sedimentary basins

Organisms under go progressive morphological changes called morphogenetic or phyletic series or lineage with variable rates depending up on the rate of change in environment over a geological period, and once changed, the morphology does not revert even if the same set of conditions are met again. Some aspects of their biostratigraphy that are used in 'source-rock evaluation' and paleoenvironmental analysis follow:

Biostratigraphic correlation: Fossils help establish stratigraphic-units based of a fact that all the sediments deposited during a particular period of time preserve comparable forms of organisms. This gave rise to the concept of 'biostratigraphic zonation' at the following levels:

Datum plane, biozone and higher-resolution biostratigraphy: The characteristics of the distribution of microfossil-assemblages should be interpreted to correlate 'wells'/ 'standard section'. A datum-plane/a boundary of biostratigraphic change of distinctive character (marker species) based on abrupt evolutionary appearance 'First Appearance Datum (FAD) or 'Last Appearance Datum (LAD) should be a conspicuous paleontological event, correlatable and isochronuous over reasonable distances.

Marker species, first recognized in Quaternary sequences worldwide, helped correlate them with the uppermost Cenozoic sequences where the method has been applied in inter-basinal and intra-basinal correlation of marine sections.

A biozone is the lowest unit of a biostratigraphic classification at a regional level for any kind of biostratigraphic unit, where 'upper'- and 'lower-limits' are defined by datum-planes. They may be defined as:

(a) Assemblage-zone *i.e.*, a body of strata whose content of fossils or certain fossils taken in its entirety, constitutes a natural assemblage or association that help distinguishing it in biostratigraphic character from adjacent strata (Fig. 3.18 A);

(b) Range-zone, represents the stratigraphic range of one species, a group of species, a lineage and may be recognised as 'Taxon-Range-Zone' (Fig. 3.18 B); 'Concurrent-Range-Zone' (D); or a 'Lineage-Zone' (E);

(c) Acme-zone' is a body of strata representing maximum development of some species, genus, or other taxon, but not its total range (Fig. 3.18 C); and

(d) Interval-zone is a gap between two distinctive biostratigraphic horizons and is not itself a range-zone of any taxon. It recognised on the basis of absence of a particular distinctive biostratigraphic assemblage or feature during the span when the rocks were deposited (Fig. 3.18 A).

Stratigraphic correlation: In the global and local-correlations, such as from 'well to well', foraminifera provide some important information. If the two wells are located in the same sedimentary basin, there are some methods for correlation between wells A and B. One is comparison of the changes of planktic and benthic foraminiferal-ratio (P/B ratio) between them. The ratio is calculated as follows:

$$\frac{\text{Planktic foraminiferal number}}{\text{Planktic foraminiferal number} + \text{benthic foraminiferal Number}} \times 100\,(\%)$$

This ratio indicates degree of Open-Ocean influence; a high value means that the whole sedimentary basin was under such influence. Because these influences affect mast (flagpole) of an entire basin in the same way, changes in the P/B ratio are useful indicators for well-to-well correlation. Another method is a ratio of arenaceous and calcareous benthic foraminifera (A/C ratio). This ratio is calculated as follow:

$$\frac{\text{Arenaceous foraminiferal number}}{\text{Total benthic foraminiferal number}} \times 100\,(\%)$$

In this case, the high value of this ratio indicates a stagnant environment in the basin.

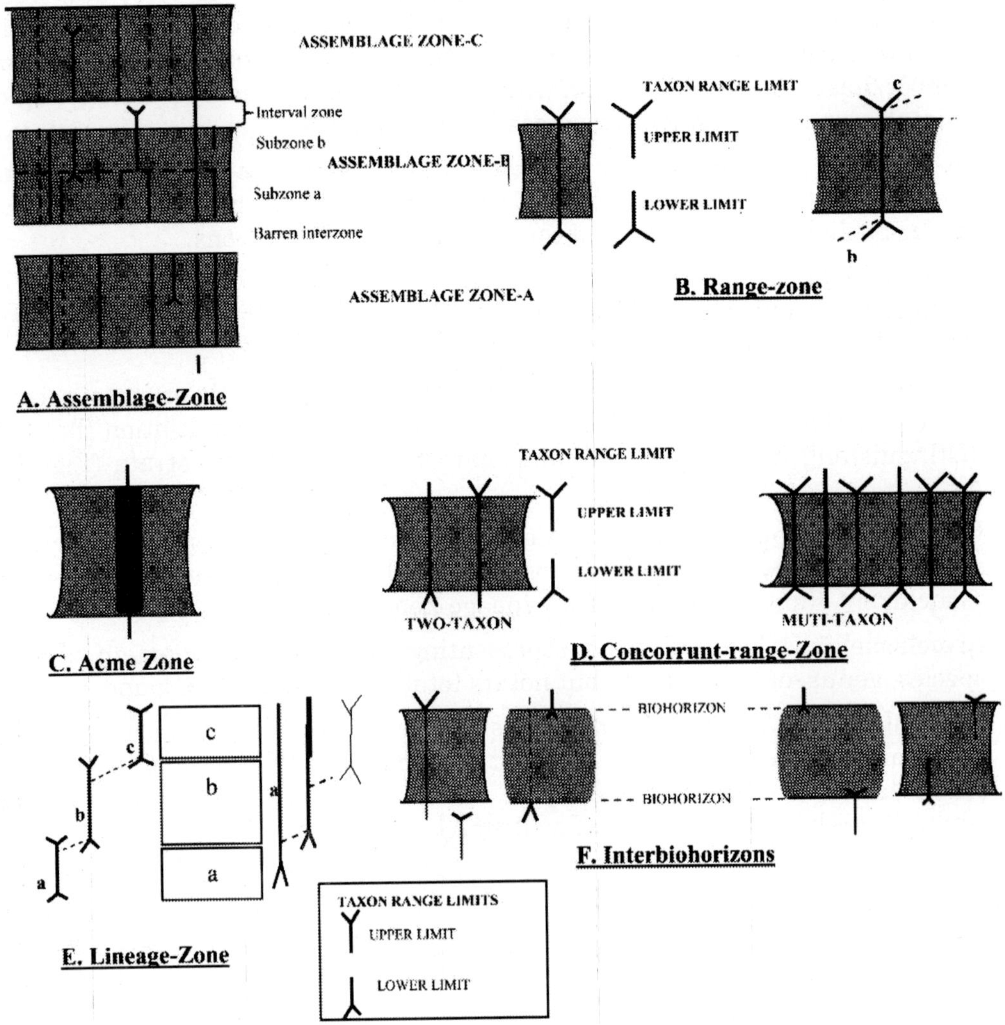

Figure 3.19 Kinds of bio-zones

Interpretation of paleoenvironment: Usually interpretation of paleoenvironment is based on microfossil characteristics. The methods of paleoenvironmental analysis are discussed as follows:

i. *Recognition of geologic phenomena:* Unconformities, sea-level changes, climate changes, etc., are usually reflected in lithofacies. The data helps studying the causes of geologic phenomena. In a general sense there are two types unconformities: subaerial erosion (reflected in distinctive lithofacies) and non-deposition or subaqueous erosion (sometimes unrecognised as a change of lithofacies). These unconformities are interpreted based on paleontological data.

ii. *Subaerial erosion* Subaerial erosion may be marked by lithofacies change between the rock below the unconformity and the basal conglomerate.

iii. As 'lithozonation' has its obvious limits, other 'on hand' criteria may help achieving better/'higher-resolution' e.g., magnetozones; seismozones; biozones and subbiozones (based on microfossils) in a an imaginary succession (Fig 3.18).

iii. *Non-deposition or subaqueous erosion*: A lot of information on deep-sea sediments has been published, for e.g., presence of deep-sea hiatuses were detected at many sites. If both the benthic foraminiferal assemblage and the lithofacies are unchanged across the unconformity it indicates that the paleo-water depth was constant. Deep-sea hiatuses are caused by changes in bottom currents. If the bottom current velocity is fast, the grains move instead of being deposited. So, the direction of bottom current is indicated by tracing the presence of the hiatus at a number of sites.

iv. *Sediment accumulation rate*: The ages of each of the datum planes in forambiostratigraphy were recently estimated by interpolation between magnetic reversals. On the basis of the depth of each datum plane in an exploration well, the accumulation rate can be calculated. Accumulation rates are related to the environment of the sedimentary basin, and a change in accumulation rate indicates a change in that environment. On the basis of five planktic foraminiferal biohorizons, the ages at each depth are plotted in a section. Sediment accumulation rates are calculated based on the thickness between biohorizons and on their ages. If the benthic faunal data are added, the cause of the environment change may be clarified. Sea-Level Change Evidence help recognising sea-level changes in the geological section. Such changes influence both the distribution of reservoir rocks and the production volume of organic matter. Changes in the volume of the mid ocean ridge system has probably produced the general rise and fall of sea-level during the Cretaceous and Tertiary 350 m sea-level change at maximum. The most rapid significant changes of sea-level are those generated by the formation and melting of ice caps. The sea-level change is estimated to be about 100m from the glacial eustatic rise and fall.

v. *Cyclothem*: In a case study of Carboniferous the fossil evidence showed a paleoenvironmental change from non marine, near-shore, off shore, and finally to near-shore within each cyclothem. Some individual beds within the cyclothems could be traced for more than 300 km. It seems likely that cyclothems are the product of eustatic sea-level fluctuations induced by repeated glaciation.

vi. *Faunal Change*: In the upper Cenozoic sediments various methods have been employed for the analysis of paleoenvironment. In a case study the sea-level fluctuations and climatic events were analysed using foraminiferal assemblages in the southern Gulf of Mexico.

2. Seismic Stratigraphy

Sedimentation, responding to changes in sea level, has produced a stratigraphic record by shaping sedimentary sequences that can be studied by seismic reflection. However, it is necessary to compare seismic data with paleontological data to determine global changes in the sea level.

3. Geohistory analysis

With the understanding of the characteristics of microfossils and how to use them to analyse geological ages and paleoenvironment a numbers paleontological properties (e.g., age in years and water depth in meters) is used in numerical stratigraphy called 'Geohistory Analysis' (Van Hinte, 1978).

3.9 STRATIGRAPHIC DISTRIBUTION OF FORAMINIFERA

The purpose of the stratigraphic distribution is to summarise the principal occurrences of foraminifera in the sedimentary rocks of the geological column. The geological range of suborders is shown in the Fig. 3.19.

The foraminifera are considered as one of the primitive forms of animal life owing to their unicellular structure, but many higher invertebrate e.g., trilobites have shown earlier record. The earliest species were similar to the present day members of suborder Allogromina having tectinous wall that could not be preserved.

Precambrian

The earliest known record of micro-organisms has been assigned to foraminifera, usually to the genus *Orbulina*. Certain objects which found in the Precambrian quartzites of France turned out to be doubtful or referred to as Radiolaria. Hence, no clear picture of Precambrian foraminifera has emerged as yet.

Cambrian and Ordovician

Jones (1956) pointed out that the foraminifera may be represented by forms resembling *Psammosphaera*, *Hyperamminoides*, *Lagena*, *Nodosaria*, and *Marginulina* but their identification is subject to question. In general the Cambrian foraminiferal fauna seems to be poor, all over the world and most of the earliest texa were structurally simple and primitive forms having single chamber or of cluster of chambers. Ammodiscaceans resembling *Bathysiphon* and *Tolypammina* have been reported from the early Cambrian rocks but they may also be interpreted as calcareous algae of similar age (Brasier, 1977).

Loeblich and Tappan (1964) opined that their original forms were doubtful owing to crushing of soft pseudochitinous or weak tests than the fauna of the succeeding periods of Ordovician and Silurian that had more number of chambers and, were rich both, in the genera and species. The allogromiid genera viz. *Chitinodendron* (U. Cam.), *Archaeochitosa* (Ord.), and *Chitinolagena, Labrinthochitina,* and *Mayliosoria* (U. Ord.) are the oldest pseudochitinous foraminifera. The oldest agglutinated forms belong to Cambrian and Ordovician Asterorhizidae, viz. *Bathysiphon* (L. Cam.), *Hyperammina* (L. Ord.), *Astrorhiza* (M. Ord.), and *Rhabdammina* (U. Ord.) and Ordovician Saccamminidae (*Ordovicina, Kerionammina, Pseudoastrorhiza, Tholosina*). The oldest known calcareous forms are also of Ordovician age and belong to Parathuramminacea of the Family Moravamminidae (*Saccamminopsis*).

Silurian

By this time wide diversification in the foraminiferal assemblages had occurred. Therefore, in addition to the Cambrian forms the allogromids was represented further by *Archaeochitina* and *Xenotheca*; Saccamminides by *Blastammina, Saccammina, Lagenammina, Thurammina,* and *Colonammina.* This period also witnessed appearance of the first enrolled tubular forms grouped in the Ammodicidae with *Ammodiscus.* Other genus like *Rhabdammina, Hyperammina, Tholosina, Thurammina, Webbinella,* and *Marsipella* also appeared in Silurian. Microgranular calcareous genera and the earliest Nodosinellidae represented by *Illigata* and *Eolagena* also appeared in Silurian, in its later phase.

Devonian

The hard-test foraminifera were scarce until the Devonian (Brasier, 1980). A few additional agglutinated genera of families Astrorhizidae (*Hyppocrepina*), Saccamminidae (*Ceratemmina, Webbinelloidea*), Ammodiscidae (*Psammonyx, Trepeilopsis*), Lituolidae and the first forms of Endothyridae and Laginidae are common during Devonian. Among the Laginids *Lagena, Cristallaria* are reported from the Devonian of Europe. This period is characterised largely by the great expansion of microgranular forms with the first appearance of Parathuramminidae (*Parathurammina, Archaesphaera, Cribrosphaeroides, Irregularia*). Family Semitextulariidae, the earliest Tournayellidae and the first Endothyridae appeared in the Devonian.

Carboniferous (Mississippian and Pennsylvanian)

This period is the most important period in the history of foraminifera, because of the world wide dominance of stratigraphically significant endothyrid limestone (*Endothyra, Plectogyra* and their various species) occurring in the Mississippian and fusulinids (e.g., *Milerella*) beds throughout the Permo-Carboniferous. A few additional agglutinated genera of Ammodicidae appeared in Carboniferous but the most significant development was recorded in the form of more complex forms, the Lituolacea, which are the earliest porcelaneous calcareous genera besides the expansion of microgranular calcareous genera especially the Endothyracea and Fusulinacea.

Some of the other important developments were the first appearance of Homosinidae (*Rheophax*), Lituolidae (*Haplophragmoides, Trochamminoides, Ammobaculites*), Textulariidae (*Textularia, Spiroplectammina*), Trochamminidae (*Trochammina*) and Ataxophragmiidae (*Mooreinella*). The a additions in porcelaneous genera in the Mississippian are the tubular, enrolled Fischerinidae (*Cyclogyra, Agathammina, Hemigordius*) and in the Pennsylvanian, the first 'chambered genus' *Eosigmoilina* (?Triassic). The Pennsylvanian fauna was dominated by the fusulinids, which were derived from an endothyrid ancestor and arose as close-coiled forms with short axis in the Upper Mississippian. They are considered as one of the excellent extinct guide fossils among the foraminifera. The Lower and Middle Pennsylvanian is characterised by fusuliniid wall structure while the most of genera of the Upper Pennsylvanian by schwagiriniid wall structure.

Permian

Fusulinids, especially the schwagiriniids first reached to their final expansion by addition of many developing long attenuated tests and highly fluted septa, and finally became extinct in Permian. The additional genera of agglutinated, microgranular and porcelaneous forms were added with internal complexities. So far the perforate hyaline calcareous foraminifera are concerned; their earliest nodosarids viz. *Nodosaria, Dentalina, Frondicularia* etc., first appeared in the Permian.

Triassic

The early Triassic fauna witnessed a fall in foraminiferal numbers. However, in the late Triassic the oldest arenaceous families, viz., Verneuilinidae, (*Gaudryina*), Valvulinidae (*Valvulina*), porcelaneous Soritidae evolved. Family Nodosariidae dominated the Triassic period by addition of the ornamented forms and certain new genera like *Lenticulina, Marginulina, Vaginulina*, etc. Additional genera in families, viz. Polymorphinidae, Bolivinitidae, Involutinidae and Spirillinidae were added. The Triassic fauna also included the first members of Anomalinidae and Discorbidae. The Triassic period is also known for the development of most of the calcareous superfamilies except Rotaliacea and Orbitoidacea.

Jurassic

During the Jurassic times, the foraminiferal fauna once again became very rich. Agglutinated forms of Lituolidae and Astrorhizidae increased in abundance with new form such as *Rheophax, Ammomarginulina, Ammobaculoides* etc., making their appearance in Jurassic. The most dominant members of the fauna, however, the Nodosariacea was represented by many genus, most of them had ornately ribbed surfaces. This period also witnessed the first attached Polymorphinidae and the Glandulinidae

Expansion of Lituolidae and Dicyclinidae also occurred during this period. Among the porcelaneous forms the earliest were- Nubeculariinae, Miliolidae (*Quinqueloculina, Triloculina and Pyrgo*) and Rhapydionininae. Lagenidae was also an important group of

Jurassic time along with the first appearance of Buliminidae and a few Rotaliidae, e.g., *Nonion* and *Discorbis*.

Cretaceous

Cretaceous is period which is unparalleled so far as the abundance of foraminifera; fauna is concerned. This period is also known for the appearance of first planktonic forms. The important amongst them are: Heterohelicidae (*Heterohelix*), Planomalinidae (*Planomalina*, *Globigerinelloides*), *Schakoinidae* (*Sakoina*), and Rotaliporidae (*Rotalipora*, *Hedbergella*).

In the middle Cretaceous they suddenly increased in the number and the variety of genera. The additional new forms developed in the Late Cretaceous are: Heterohelicidae (*Gublerina*, *Planoglobulina*, *Pseudotextularia*), and Globotruncanidae (*Globotruncana, Rugoglobigerina*). Besides, the earliest Globigerinidae (Globorotaloides) appeared in latest Maastrichtian. The most important development in Cretaceous was the formation of foraminiferal limestones, the lower part of which includes miliolids, and the upper orbitolinid, alveolinid and orbitoidid. This period is also known for the local evolution of agglutinated forms. Some the other agglutinated genera had labyrinthic interior. (e. g., *Cribratina* and Hormosinidae). The first Rzehakinids (*Rzehakina, Miliammia*) are found in Cretaceous beds along with many new Lituolids (*Ammobaculoides, Cyclammina, Spirocylina, Loftusia*). Similarly Ataxophragmiids, Discocyclinids and Orbitolinids were also expanded by their representative viz., *Orbitolina, Dictyoconus* etc.

The limestones of Late Cretaceous have abundance of imperforate calcareous forms represented by the earliest Spiroloculininae (Spiroloculina) and many genera of Miliolidae, additional Soritidae and the earliest Alveolinids. The hyaline calcareous forms of the Cretaceous times, however, are characterised by a great expansion of coiled genera representing most of the superfamilies. In Buliminids a new forms of the Bolivinitids are *Bolivina, Bolivinoides, Tapanina* etc.; in Turrilinids new forms of Turrilinidae (*Buliminella, Neobulimina, Pyramidina, Rectobulimina, Spirobulimina* etc.) In the Uvigerinids the earliest genera are Late Cretaceous *Orthokarstenia* and *Pseudouvigerina*. New Discorbids are *Conorbina, Epistominella, Baggina* etc. The oldest Patellinids occur in Lower Cretaceous rocks; also the oldest Rotaliacea, represented by Rotaliidae (*Rotalia, Kathina, Pararotalia*), Calcarinidae (*Calcarina*) and Nummulitidae (*Selcoperculina*). The Orbitoidacea particularly Cibicididae (*Cibicides, Planulina*), Cymbaloporidae (*Cymbalopora*) and Orbitoididae (*Orbitoides*) entered the fossil record in the Late Cretaceous. *Fursenkoina, Loxostomum, Globorotalites, Chilostomella, Nonion, Nonio-nella* and *Pullenia* and a few Involutinidae also occur from the Late Cretaceous strata.

Paleocene

The species found show transitional assemblages including some elements of the late Cretaceous (Maastrichtian). The larger foraminifera added were the porcelaneous *Faciolites* and *Orbitolites* together with the earliest *Nummulites* and *Discocyclina*. Among smaller taxa, new nodosariacean genera (*Ploymorphina, Glandulina*), and buliminacean genera (the earliest Buliminidae, *Bulimina, Globobulimina*) appeared.

Besides, new Rotaliacea (first Elphididae, *Elphidium, Protelphidium*) and a new Rotaliidae, granular-walled *Florilus, Gyroidina, Melonis* were also added. The important change were noted in the planktic families of Cretaceous Rotaliporidae, Globotruncanidae and Schakoinidae which disappeared at the close of Maastrichtian, as did nearly all the Heterohelicidae, except *Heterohelix* and *Bifarina*. The Globigerinidae established it as the dominating among all the planktic families in Paleocene. The important species were *Globigerina, Globoconusa, Globorotalia* etc.

Eocene

The period was dominated by both, larger and smaller forms. Orbitoides and other larger species of the reef environment expanded considerably. Nummulitidae became so abundant, that 'Nummulitic limestone' became a synonym to Palaeogene in the European Stratigraphy. Alveolinid limestones also formed in early and middle Eocene. Lepidocyclinids (*Lepidocyclina*), ataxophragmiids (*Clavulina*), miliolids (*Haurina, Articulina*), nodosariids (*Amphimorphina,*), buliminids, uvigerinids (*Uvigerina*), and pavonids *(Reusella)* appeared in Eocene. New genus of discorbids (*Discorbis, Cancris*), glabratellids (*Glabratella*), siphoninids (*Siphonides*) and epistomariids are found in Eocene fauna. New rotaliids of Eocene are: *Dictyoconides, Elphidium, Eponides, Planorbulina, Cassidulina* and *Pseudobulimina*. Agglutinated forms- astrorhizids and lituolids became extinct in the Eocene, e.g., *Martinotinella* and *Lituonella*. The important planktic genera of the Eocene are, namely- *Globigerinoides, Globoquadrina* and *Globigerapsis*.

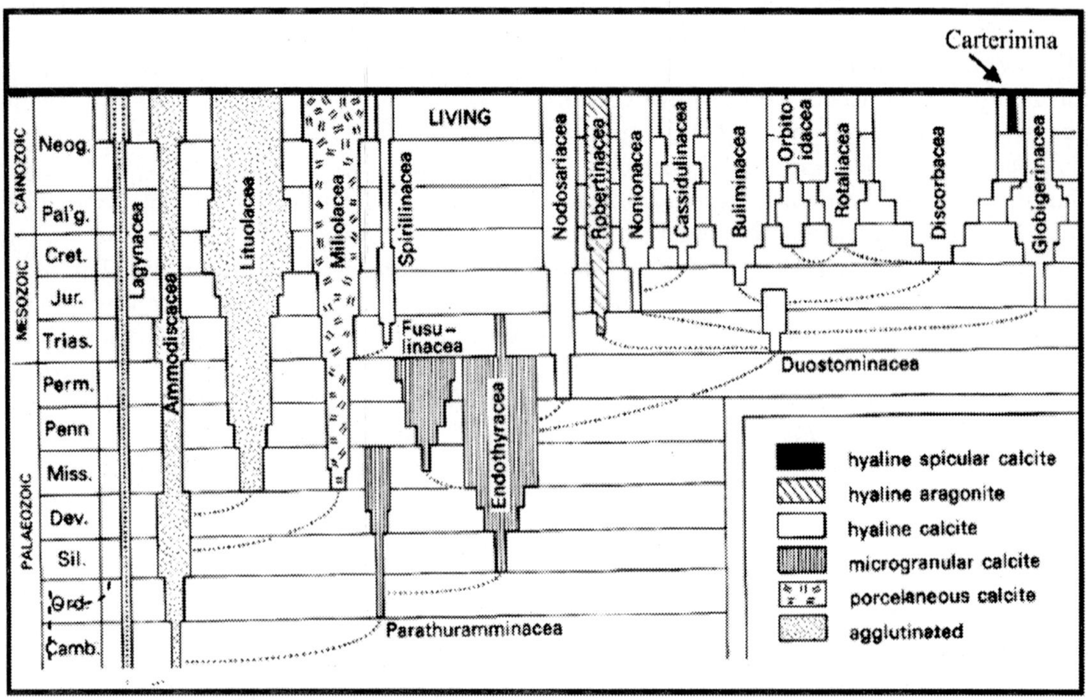

Figure 3.20 The phylogenetic history of the Foraminiferida (after Braiser, 1980)

Oligocene and Miocene

The larger species are of particular importance in Oligocene and early Miocene strata. Some that proved to be very good index and guide fossils are Miogypsinidae (*Miogypsina* and *Miogypsinoides*) at global level and Lepidocyclinidae at local level. Some of very complex Soritidae and some Rotaliidae, Bolivinitidae, Calcarinidae (*Baculogypsina*) and Anomalinidae (*Hanzawaia*) are also among the abundant fauna of Eocene.

Pliocene and Pleistocene

The fauna of Pliocene and Pleistocene, in general are characterised by the increasing abundance of planktics forms and decreasing abundance of larger species. The important appearances of taxa in Pliocene are *Pseudorotalia*, *Geminospira*, *Caribeanella* and *Pulleniatina*. In the Pleistocene the important additions are *Asterorotalia* and *Hyalinea*.

3.10 LARGER FORAMINIFERA

a. Introduction and habitat

They include a variety of forms, which are not related taxonomically. Many are extinct but are good index fossils of the Tertiary, particularly for the Eocene times.

They are distinguished from the smaller benthic foraminifera by their larger size (more than 1 mm) and complex endoskeleton features. As a result, they are studied in thin sections (equatorial; axial and tangential) whereas the smaller ones are studied as whole specimens. The larger size of the individuals does not appear to restrict locomotion, and specimens up to 3 cm in diameter are known to move around on the substrate or on vegetation for feeding on diatoms and algae. Some larger foraminifera ingest photosynthetic algae (zooxanthellae), which then live within the cytoplasm of the foraminifera in a symbiont relationship.

b. Composition of test and arrangement of chambers

The composition and structure of the walls of larger foraminifera are important. Agglutinated, microgranular, perforate and imperforate walls are known: (i) The agglutinated matter is held together by organic cement in textulariids and (ii) Microgranular walls of fusulinids are simple or multilayered. Fusulinids are all multi-chambered. Growth starts from an initial chamber called the proloculus and chambers are added to form the coiled shell by extension of the spiral wall. *Fusulina* is a good example of Upp. Carboniferous-Permian); (iii) Larger porcelaneous forms fall in the family Alveolinidae and family Soritidae (e.g., *Orbitolites*). The alveolinid wall is imperforate, porcelain-like and bi-layered. The nummulites and orbitoids have perforated calcareous tests; and (iv) The nummulitids are commonly known as 'petrified lenses' or coin stones. The planispiral tests are lens/disc shaped and multi-chambered with a finely branching system of canals. The tests are calcareous and perforated.

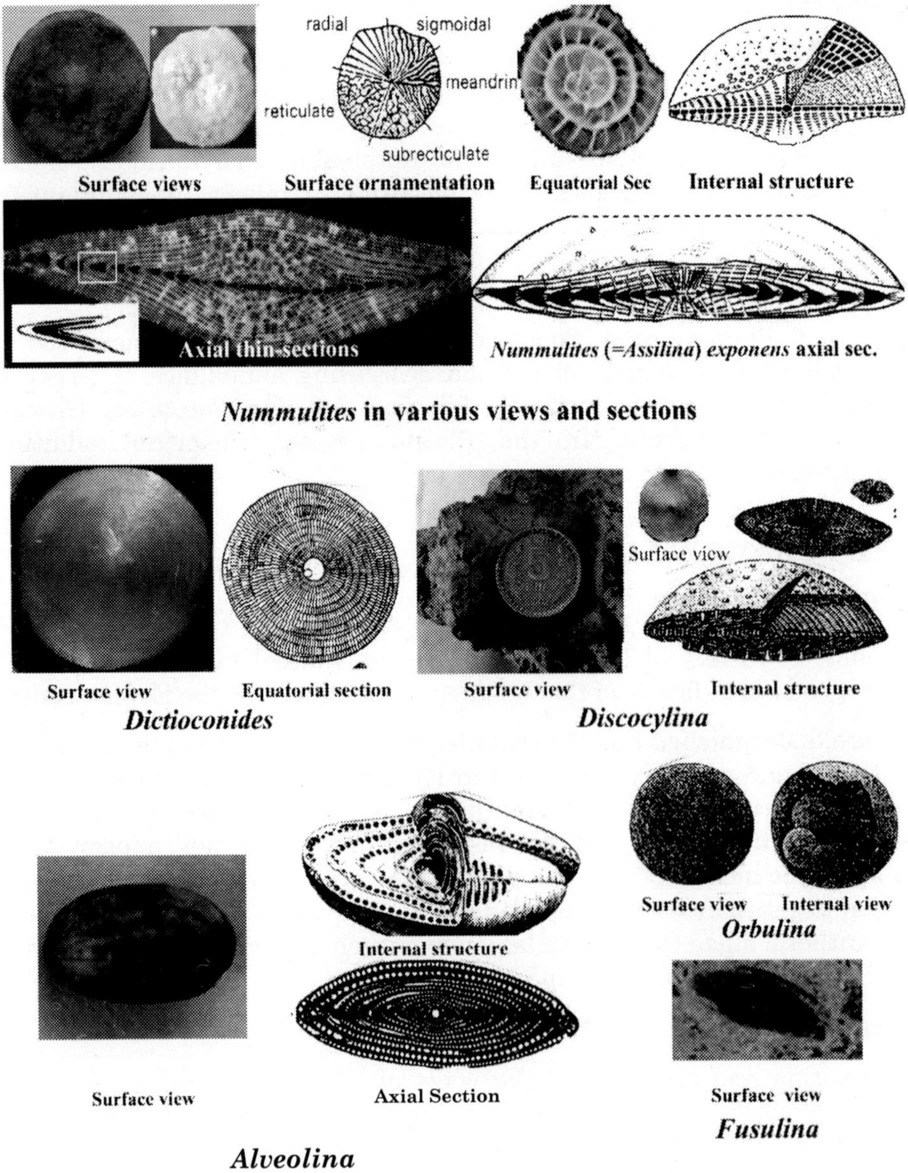

Figure 3.21 Some larger foraminifera (coin-diameter 13 mm)

Most of them are planispiral and are either evolute or involute. Some, however, are uniserial as in some textulariids (*Orbitolina*, which has uniserial-stacks of saucer-shaped chambers following an early trochospiral stage- Low.-Upp. Cretaceous). The tests of orbitoids are radial, hyaline and perforate with a discoidal growth-pattern, i.e., the chambers are arranged in rings. In thin sections the median-layer of equatorial-chambers may be differentiated from the lateral-chambers (e.g. *Discocyclina*). They have large, disc-, saddle- shaped or biconcave with a single-layered wall. The chambers are arranged in circles in equatorial-plane, surrounded by many layers of lateral-chambers.

c. Embryonic structure

In megalospheric forms, the embryonal shell (nucleoconch) consists of 2 chambers, the 'protoconch' and 'deuteroconch', or several chambers. The nucleoconch is surrounded by 'periembryonal' chambers. The equatorial chambers are connected to each other by stolons (tubes). In the layer of lateral chambers, buttresses are often formed which appear as nodules at the surface. When these concentrate in the polar region, they can give rise to a polar-plug. They range from Campanian to Maastrichtian (e.g., *Orbitoides* and *Omphalocyclus*). The genus *Discocyclina* ranges from early to middle Eocene.

d. Ecology of larger foraminifera

Some of the recent trends of interpretations of the ecology of larger foraminifera from India are enumerated as follows:

(a) Ghosh *et al.* (2008) studied intertidal foraminifera in the macrotidal estuaries of the Gulf of Cambay: Implications for interpreting sea level change in palaeo-estuaries.

(b) Saraswati P.K. (2007) gave a general account of the symbiont-bearing benthic foraminifera of Lakshadweep.

(c) Muthikrishnan and Saraswati (2001) made an interpretation of taxonomic based on shape-analysis of the nucleoconch of *Lepidocyclina* from Kutch.

(d) Saraswati and Patra (2000) used multivariate analysis of three closely resembling species of *Nummulites* to interpret the Middle Eocene of India.

e. Geological history

They appeared during late Palaeozoic. The calcareous, non-laminar, microgranular types (e.g., fusulinids) appeared first. During late Cretaceous, they witnessed their major development and forms like *Orbitolina* and *Orbitoides* appeared. The start of Cenozoic witnessed the sudden extinction of many genera (e.g., *Orbitoides*) and emergence of groups like Nummulitidae, Miogypsinidae, Rotaliidae, etc.

The close of Oligocene brought sharp decline in the Alveolinidae, Nummulitidae and the Lituolidae.

In Recent times, the larger foraminifera are represented by no more than a few genera but the smaller are still present found living in large number, in general in only in the warm, shallow waters of the tropics and subtropics, associated with carbonates and calcareous clays. The benthic groups are characterized by the free-swimming larval stages which widely disperse them to considerable geographical distances. Biostratigraphic sequences were re-calibrated using the '*Miogypsina* Scale' in Oligocene-Mid. Miocene to achive 'high-resolution'. It has been also used in Oligocene-Miocene petroliferous rocks in Gujrat, Bombay Offshore, Kerala, Cauvery-Godavari, Mahanadi, Bengal Fan and Andaman and Nicobar basins where miogypsiniides show widespresd occurrences.

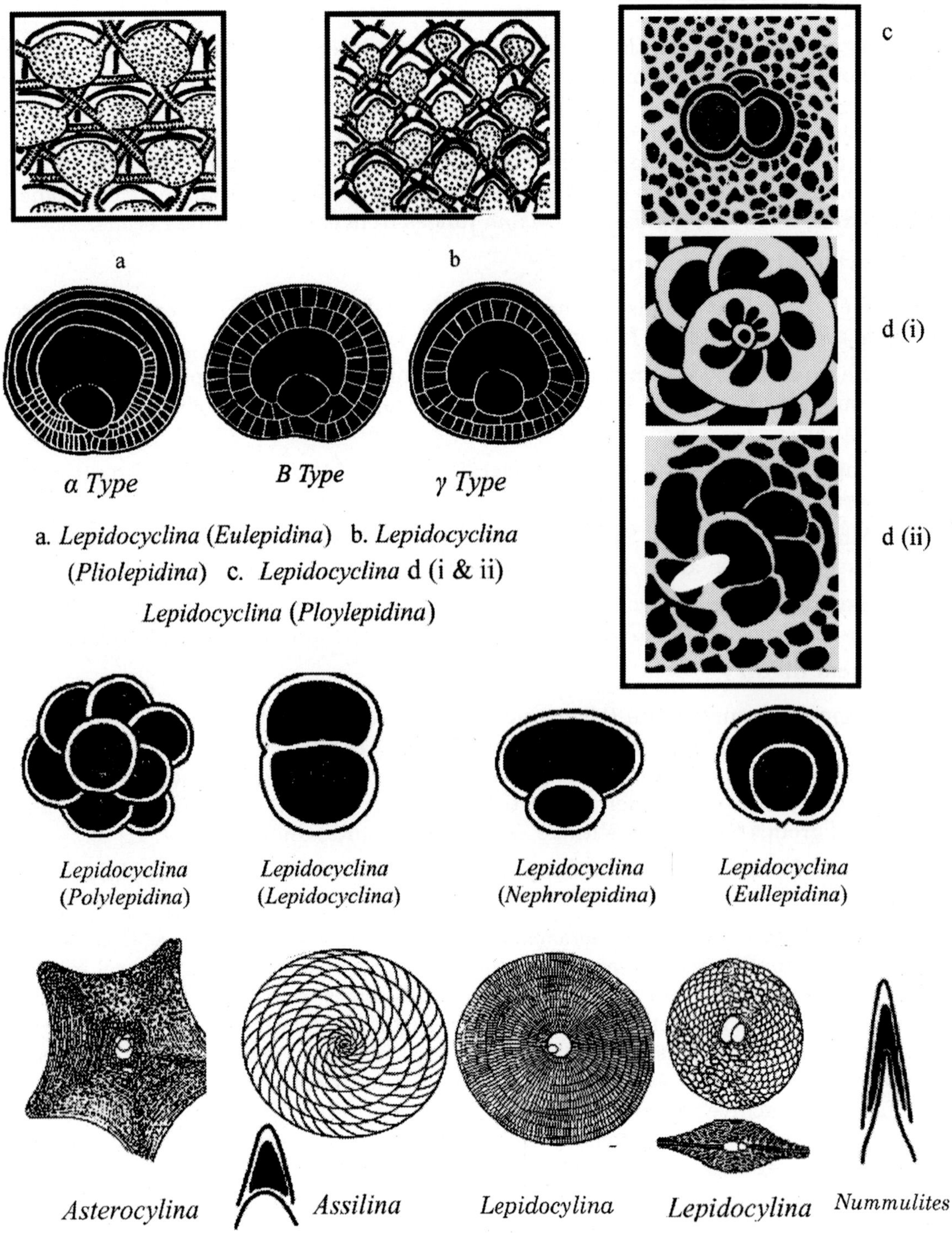

Figure 3.22 Larger Tertiary Foraminifera and their embryonic apparatus

Table 3.7 Foraminifera as index fossils (based on Loeblich & Tappan, 1988)

Age	Index Fossil
Holocene	*Gyroidina*
Pliocene and Pleistocene	*Pseudorotalia, Geminospira, Caribeanella* & *Pulleniatina Asterorotalia, Hyalinea*
Oligocene &	*Miogypsina, Miogypsinoides, Baculogypsina, Hanzawaia*
M. Miocene	*Globoquadrina, Globigerapsis*
L. Miocene	*Globigerinatella*
U. Eocene	*Asterocylina, Helicostegina, Pseudophragmina*
L. Eocene	*Dictyoconides* (in India), *Faciolite, Saccamminoides Orbitolites*
Palaeocene	*Discocyclina* (M. Palaeo.-U. Eco.), *Globoconusa, Melonis, Bifarina*
U. Cretaceous	*Cyclolina, Cyclopsinella, Gublerina, Heterohelix Loftusia, Orbitoides, Planoglobulin, Pseudotextularia, Spirocylina, Vidalina*
Cretaceous	*Ammobaculoides, Abathomphalus, Orbitolina, Regoglobigerina, Conorbina, Spiroplectinata, Ventelabrella, Neobulimina, Rectobulimina, Orthokarstenia, Pseudouvigerina, Planomalina, Globigerinelloides, Rotalipora, Hedbergella, Cribropleurostomella, Globotruncana, Rugoglobigerina, Globorotaloides*
U. Jurassic	*Nubeculinella*
U. Triassic	*Asymmetrina, Plagiostomella*
M. Triassic	*Deustomina*
U. Permian	*Fusulina, Lunucammina, Neoswagirina*
M. Permian	*Cancellina, Miscellina, Neoswagirina, Pseudodoliolina, Verbeekina, Wedekindellina*
Permian	*Cribrigenerina, Swagirina*
U. Carboniferous	*Bradyina, Fusulina, Fusulinella, Profusulinell, Cribrostomum, Haplophragmella*
L. Carboniferous	*Spiroplectammina Cyclogyra, Agathammina, Hemigordius, Earlandinita*
U. Devonian	*Colaniella*
L. Devonian	*Ceratemmina, Cribrosphaeroides*
U. Silurian	*Illigata* and *Eolagena*
L. Silurian	*Archaeochitina, Blastammina, Colonammina*
U. Ordovician	*Chitinolagena, Labrinthochitina, Mayliosoria, Webbinelloidea*
M. Ordovician	*Saccamminopsis*
L. Ordovician	*Ordovicina, Pseudoastrorhiza, Tholosina*
Ordovician	*Archaeochitosa*
L. Cambrian	*Hyperammina, Astrorhiza, Rhabdammina*
Cambrian	*Bathysiphon, Tolypammina,Psammosphaera, Hyperamminoides, Bathysiphon*
Precambrian	*Orlubila* (?)

Glossary of Selected Morphological Terms

Aborescent test. It has branches like a tree e.g. *Rhabdammina* (Fig. 3.8 f).

Accessory apertures. It is a test opening that do not lead directly into primary chambers. e.g., bullae or tagilla of planktic forms.

Agglutinated test. It is composed of foreign particles cemented together on an organic lining.

Alar projections. Wing-like extensions of test.

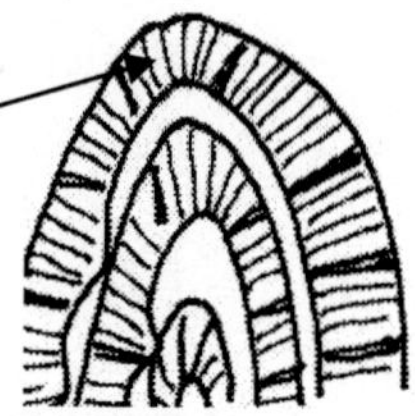

Alveolus (or alveoli) are tiny cavities in shell wall, which do not open in the test e.g., **keriotheca** in some fusulinids. Or blind chamberlets that open opposite to direction of coiling e.g., Family Alveolinidae.

Annular. Circular or cyclic or ring-like, e.g.

Areal aperture. Aperture in face of final chamber of the test.

Arenaceous test. An agglutinated test of only sand particles cemented on an organic lining.

Argillaceous test. An agglutinated test of only clay particles cemented on organic lining.

Axis of coiling. An imaginary perpendicular to the plane of coil or the line along which coiling has taken place.

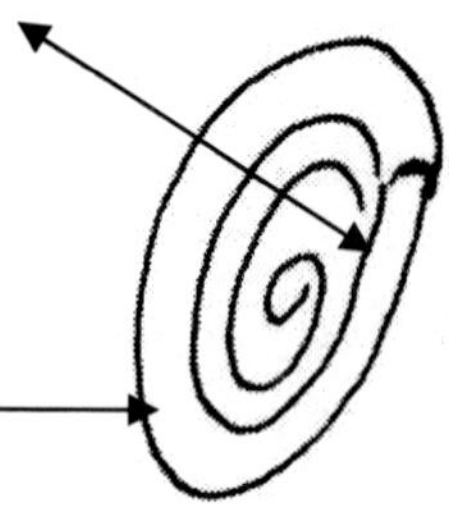

Biumbonate. A coiled test having the umbilical area raised in the form of a boss (knob) on both the sides.

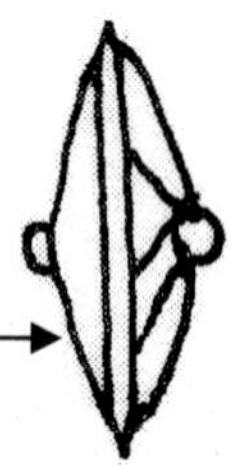

Bifid. Divided into two branches, e.g., a bifid tooth.

Bilamellar. Walls of each chamber consisting of two primarily formed layers.

Biserial test. Chambers arranged in two series

Biumbilicate. A coiled test having the umbilical region depressed on both the side.

Cameral aperture. An opening of a chamber.

Cancellate. Honeycomb pattern of ornamented test.

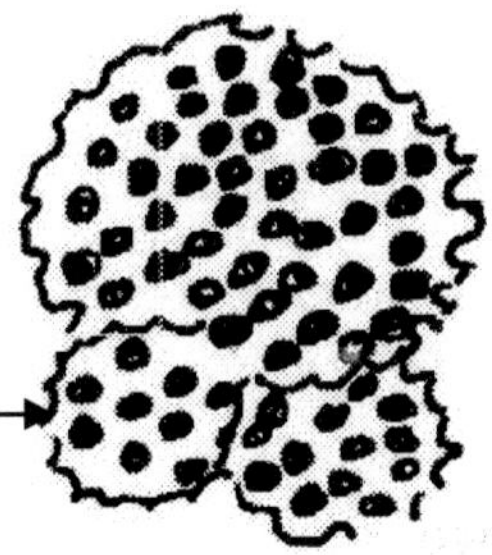

Carina. A keel or flange or carinal band or imperforate marginal area (poreless margin) e.g. *Globotruncana.*

Chamber. The test cavity and its surrounding wall, formed in a single growth stage in a multilocular test.

Chamberlet. A subdivision of a chamber produced by axial or transverse septula.

Chitin. A horny substance occurring in a few invertebrates e.g., Arthropod, erroneously reported in Foraminifera.

Clavate. Club-shaped, inflated terminally.

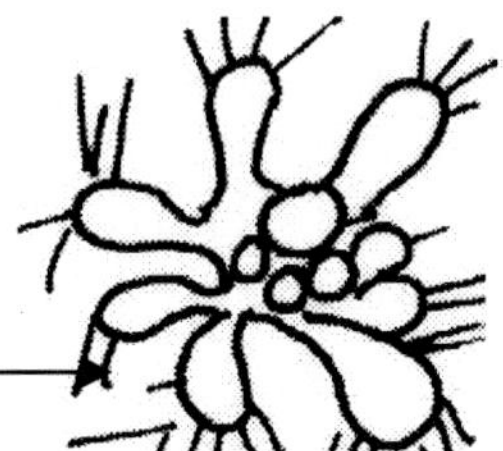

Convolute. Evolute enrolled, referring to a test with all whorls visible on both sides.

Costae. A type of ornamentation in the form of raised ridges.

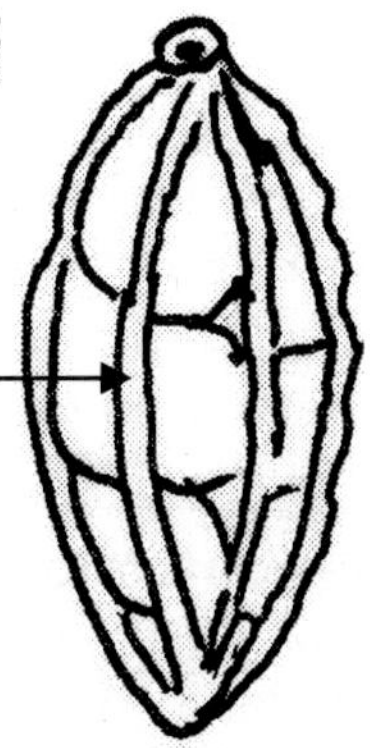

Cribrate. Sieve-like apertures on a chamber.

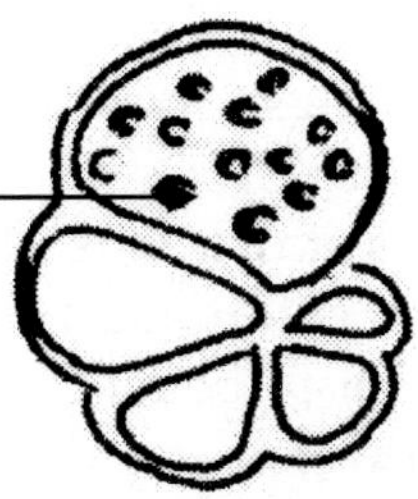

Deuteroconch. Counter-part of 'embryonic apparatus' (**neucleo-conch**) that consists of first formed '**proteroconch** and a later formed 'deuteroconch covering the former e.g. Discocyclinidae.

Deuteroforamen. An aperture in addition to the primary aperture, situated adjacent to it.

Diaphanotheca. Relatively thick, light-coloured to transparent layer of spirothecal wall next below tectum in fusulinid.

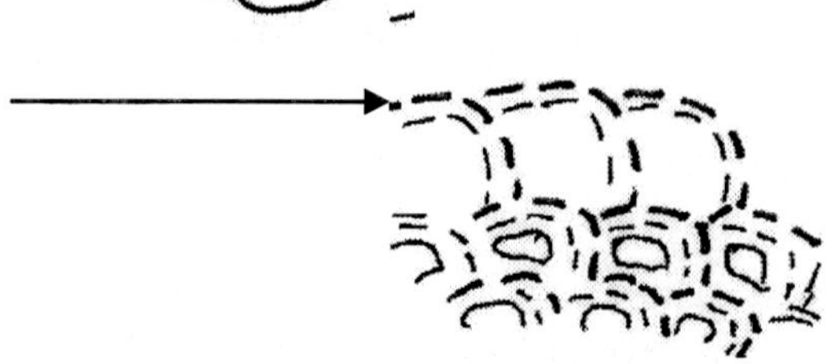

Dimorphism. Occurrence in single species two different forms; megalospheric and microspheric tests (gamont or schizont generations) Fig. 2.8.

Diploid. Stage in life history in which nuclei contains full number of chromosomes; **gamont** containing half of the number. Fusion of **gametes** forms zygote to regain diploid stage of schizont.

Distal end. The end of the test, in direction away from proloculous.

Dorsal. Spiral side of trochospiral forms (opposite to **ventral**) showing all whorls & their chambers (Fig. 3.7, *Discorbis*).

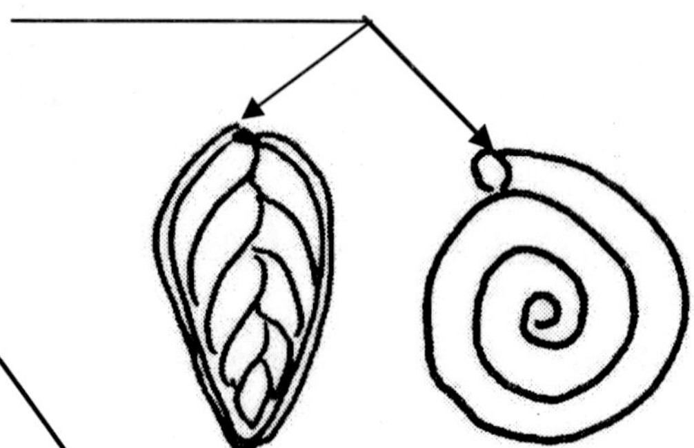

Embryonic apparatus. Group of chambers at the centre of some **megalospheric** tests, larger in size & different in shape and arrangement from other chambers; neucleoconch e.g., Discocylinidae.

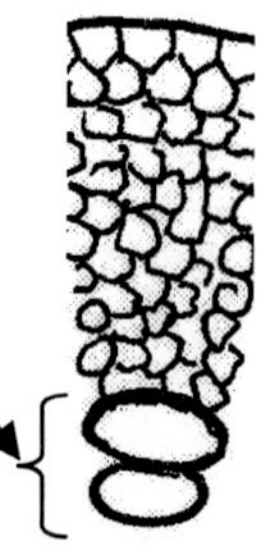

Endosymbiont. Marine algae living just below surface of shell of is called 'symbiont' that is not a true symbiosis but chloroplast sequestering allows them to benefit directly from photosynthesis

Epitheca. Secondary deposit in the inner wall of some fusulinids; tectorium.

Evolute. Tending to uncoil; chambers non embracing in coiled tests (Fig. 3.7).

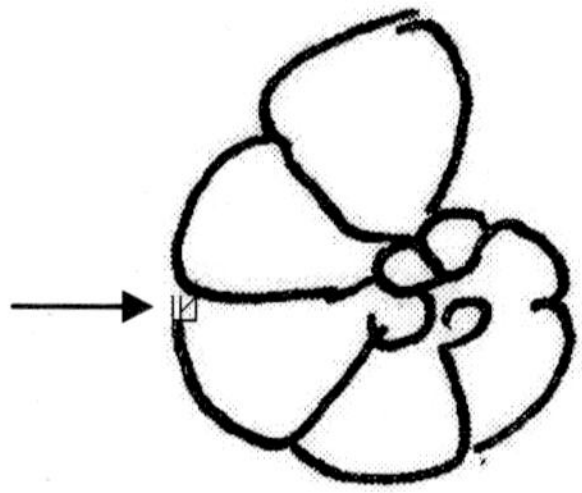

Extraumbilical aperture. Opening in final chamber of test not connecting with umbilicus, commonly sutural mid-way between umbilicus and periphery.

Flabelliform. Shaped like fan, e.g., *Neoflabellina.*

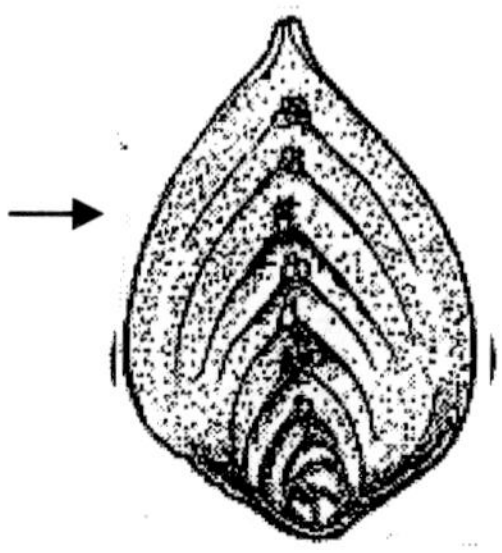

Foramen (pl. **foramina**). Opening between chambers located at the base of septa e.g., fusulinids or areal in position; in various Foraminifera may represent previous aperture or be formed secondarily.

Fusiform. Spindle shaped tapering at each end.

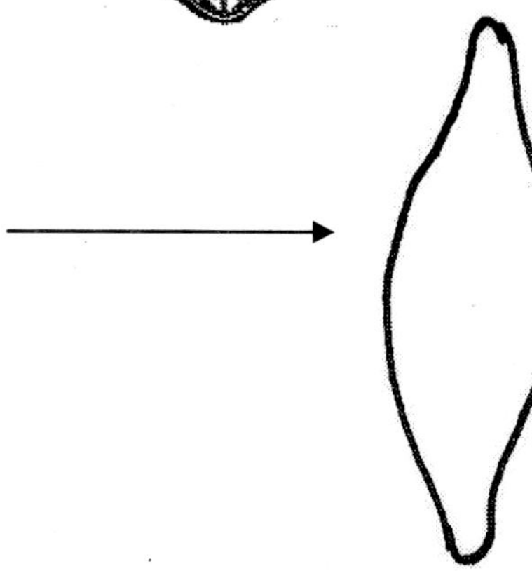

Gamete. Reproductive cell with **haploid** chromosomes number, capable of fusing in pair to form a new individual. (zygote); observed in Foraminifera are amoeboid gametes, biflagellate gametes, each chara-cterristic of distinct group of genera (Fig. 3.2).

Gamont. Generation which forms **gametes** in sexual reproduction, commonly with **megalospheric** test (A-forms) Fig. 3.7.

Granular hyaline wall. Perforate lamellar part of test composed of minutely granular calcite, seen between cross nicols as multitude of tiny flacks of colour; granular equidimensio-nal, variously oriented.

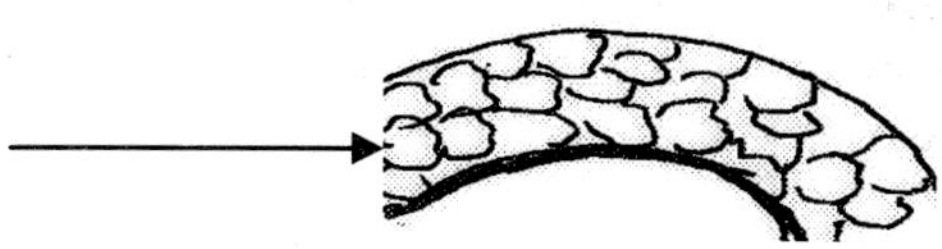

Haploid. Stage in life history in which nuclei are only one-half of the number of chromosomes; resulting from reduction division in nucleus, may be temporary in **gametes** (or only in gamet) or represent a distinct generation in Foraminifera.

Hispid. Surface covered by tiny, hair like spines, *Uvigerina*.

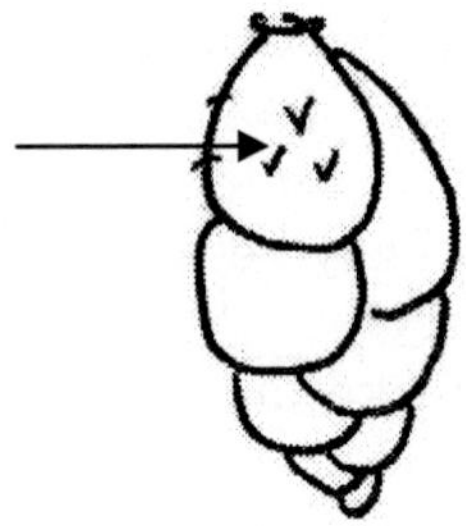

Hyaline. Glassy, transparent

Imperforate. Pores absent e.g., **porcelaneous** walls in miliolids.

Interiomarginal aperture. An aperture (opening) at the base of final chamber, along the final suture; its position in coiled forms may be 'equatorial', 'umbilical' or 'extraumbilical'.

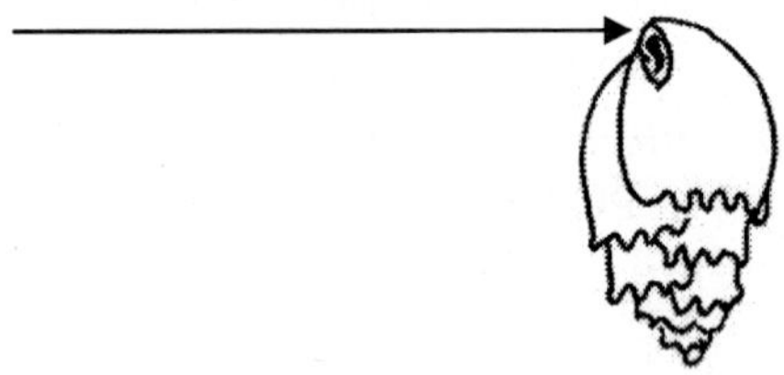

Interseptal. Located between septa (wall separating two chambers in succession).

Keriotheca. Thick layer of shell material displaying honeycomb structure, e.g., fusulinid; may consist of two layers.

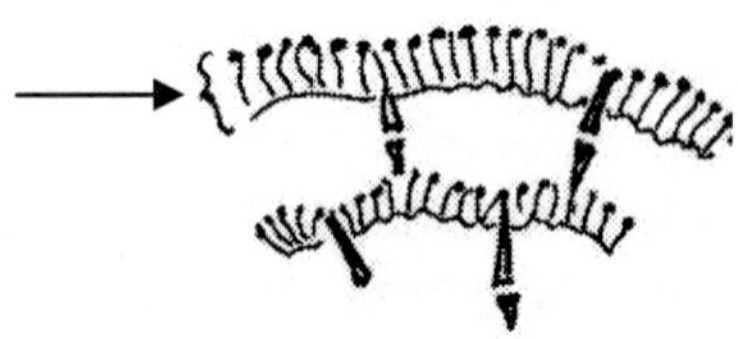

Labyrinthic. Complex spongy wall with dendritic channels normal to the surface; e.g., some agglutinated forms; (related terms: alveolar, vacuolar, vesicular).

Lamellar. Calcareous test having a of wall with more than one layer.

Lenticuline. Lens-shaped, e.g., *Lenticulina*.

Limbate. The thickened border of chamber, at sutures. May be elevated.

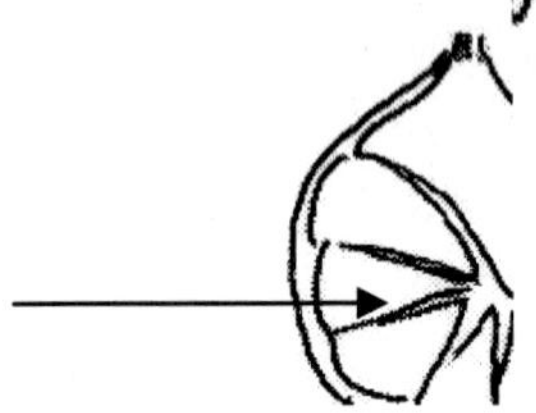

Lip. Elevated margin or border along /around an aperture.

Marginal cord. Thick spiral structure beneath surface at periphery of test, e.g., *Nummulites.*

Megalospheric. Gamont generation characterized by larger proloculus but commonly smaller test size than a test of microspheric generation.

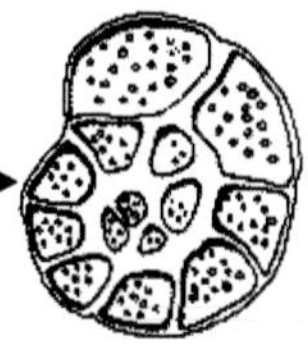

Microspheric. A schizont (agamont) generation characterised by smaller proloculus but larger test size than a test of megalospheric generation.

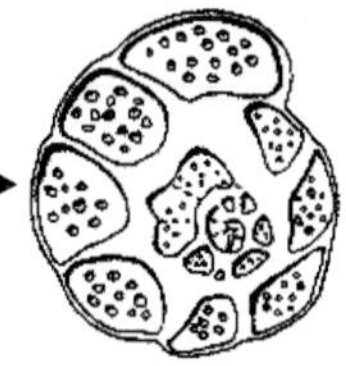

Microgranular. A wall structure composed of tiny crystals of calcite; or possibly a recrystallized form of granular a wall.

Milioline. Like miliolids, having tube-shaped chambers; two per whorl; added at a particular angles (180°, 120° or 144°).

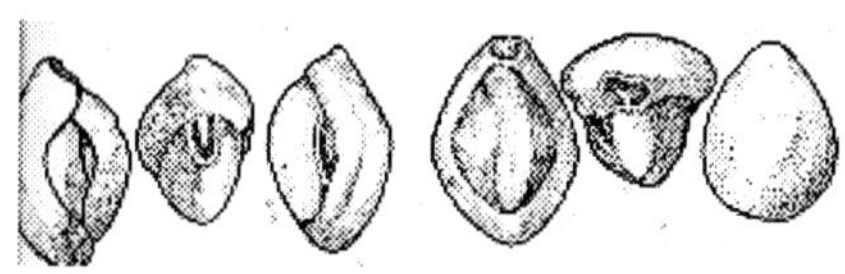

Multilocular. A test with many chambers.

Palmate. Leaf-like, flat resembling a palm with outspread fingers, e.g. *Frondicularia.*

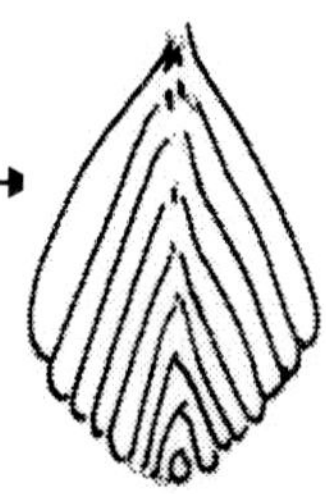

Perforate. Porous, punctate types of walls; characteristic of hyaline calcareous walls.

Phialine lip. Extended and elevated border around a neck on which the aperture is situated (like neck of a bottle) e.g. *Lagena.*

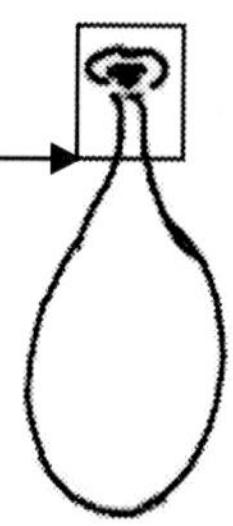

Planispiral. Coiling along a perfect planer surface.

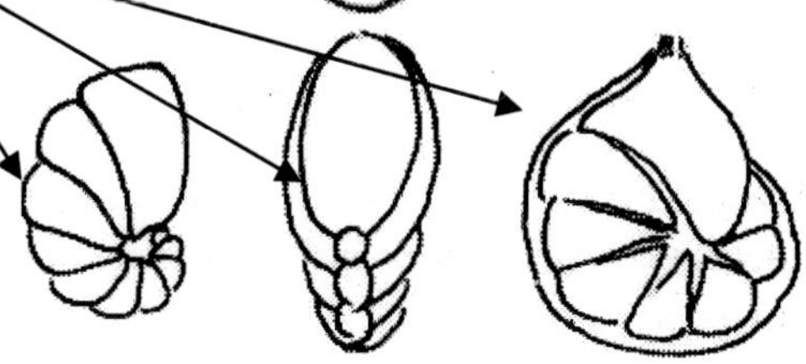

Porcelaneous. Imperforate calcareous walls, shinning like a porcelain surface.

Primary aperture. The main opening of the test; the term is used when other smaller apertures are also present.

Proloculus (pl. **proloculae**). The first formed chamber.

Quinqueloculine. A type of milioline arrangement, in which the whorls, consisting of two tubular chambers are arranged along five planes 72° apart (added at 144°); five chambers remain visible externally.

Radial microstructure. Pattern of arrangement of calcite/aragonite crystals whose c-axis remain perpendicular to the surface of the test.

Radiate aperture. An opening surrounded by a set of slits, diverging from the each other. e.g., *Nodosaria, Lenticulina.*

Septa. Partitions between two adjacent chambers.

Spinose. A surface ornamented by small spines, each comprising a single calcite crystal.

Spirotheca. Outer or upper wall of test in Fusulinina.

Stolon. Tube like extension of body connecting chambers in orbitoids.

Streptospiral. Coiled like a ball of rope.

Supplementary aperture. Openings in addition to the primary aperture.

Sutures. External impression of septa (wall) separating two chambers or spires (coils), also known as 'chamberal sutures' or 'spiral sutures' respectively.

Synecology. Observations on the ecological aspects of a community of species.

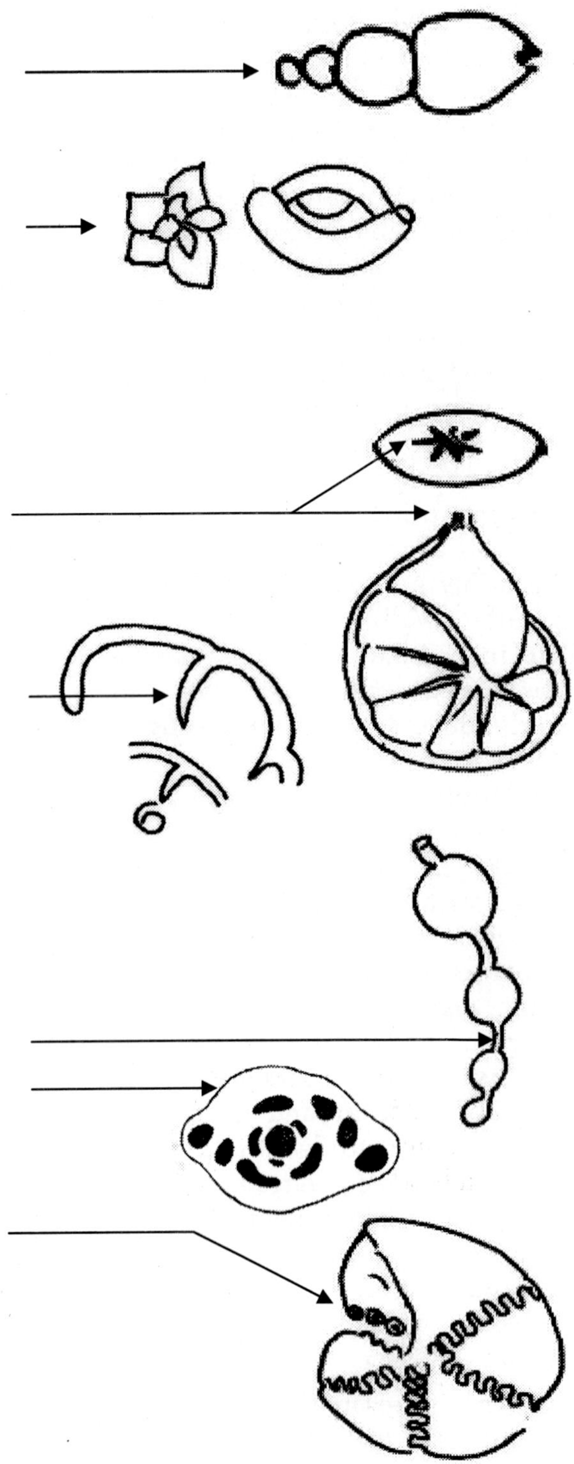

Test. Cover or shell skeleton in Foraminiferida, which may be pseudochitinous, calcareous, siliceous or agglutinated.

Tectorium (pl. **tectoria).** Internal lining of chamber, composed of dense calcite formed at or near same time as that in which tunnel in test is excavated e.g., fusulinids may include lower or upper tectoria.

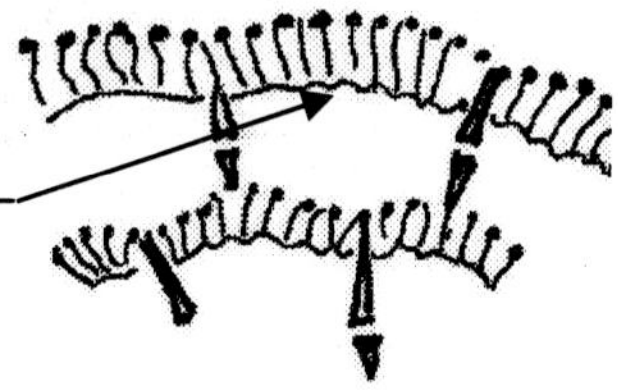

Triloculine. An arrangement chambers in miliolids where tube-shaped chambers are added at 120 degree, as a result of which only three remain visible externally, e.g., *Triloculina.*

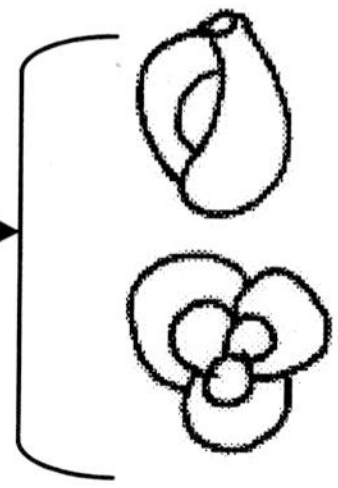

Triserial. Chambers arranged in three parallel columns, or high trochospiral with three chambers per whorl, e.g. *Verneuilina* (Fig. 3.10, c & d).

Trochoid. Trochospiral, rotaloid, rotaliform; chambers coiled spirally, evolute on dorsal or spiral and involute on ventral or umbilical side.

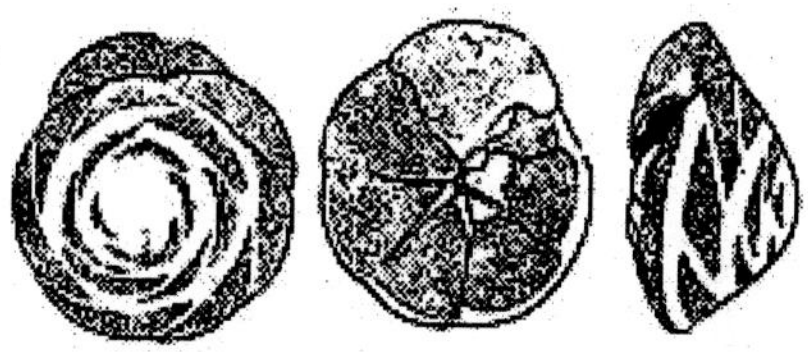

Umbilicus. Space formed between inner margins of umbilical walls of chambers belonging to same whorl.

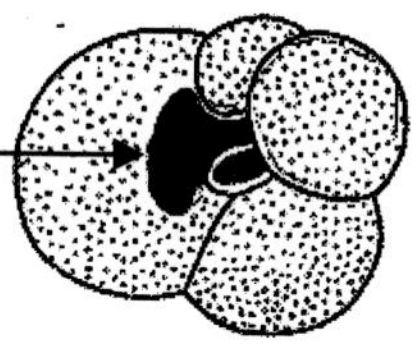

Umbonate. Having umbo on one or both sides, in the later, known as biumbonate e.g. *Amphistegina.*

Ventral or umbilical side. Pertaining to the inferior side of a trochoid test or (opposite to dorsal) in which only the outer whorl is found visible.

Whorl. Single turn or volution of coiled test (through 360 degree).

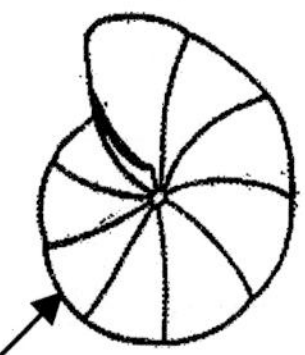

CHAPTER 4

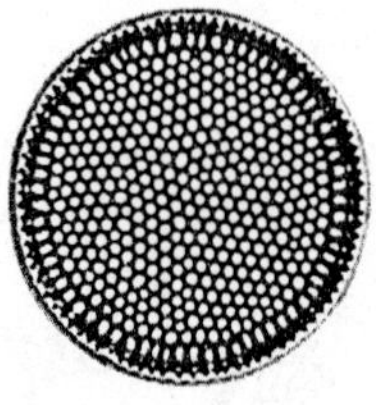

Diatoms

Contents

4.1 INTRODUCTION

Diatoms are unicellular-algae with chrysophyte-like photosynthetic pigments. They belong to Class Basillariophycae of Division Chrysophyta that includes endoplasmic-cysts forming algae with a bipartite cell wall, which secretes silica at some stage of their life cycle. Basillariophycae differs from other Chrysophytes in lacking flagella.

They commonly range in diameter/length between 20-200 μm in, but some species may grow up to 2 mm in length. The cell wall is silicified to form frustule (siliceous skeleton), comprising of two bivalves one overlapping other like a lid of a box Fig. 4.1a).

They occur both as benthos and planktons in almost every aquatic environments including the moist-soils. Generally solitary by habitat their colonial species in the form of long-chains are attached by mucous-filaments or by bands may occur in such an abundance that they can make the rock called diatomites consisting of 90% of diatom frustules. A high percentage of their living genera and make them valuable tools in paleoenvironmental studies. Deep-sea history researches show that they may be used in biostratigraphic zonations.

4.2 HISTORY OF STUDY

They have been studied since the late eighteenth century but advanced came in this field in the early nineteenth century with the emerging developments in microscopy. European workers namely, Cleve, Ehrenberg, Grunow, Schmidt and Van Heurck produced their hand illustrated monographs on diatoms in the late nineteenth century. In the early twentieth century Hustedt (1930) prepared their first taxonomic and ecological account that proved to be a key reference today. Round *et al.* (1990) published a comprehensive work on diatoms.

4.3 TECHNIQUES OF STUDY

Preparation of slides

Diatoms slides may be easily prepared and can be viewed under a light microscope. Wet sediment-/bottom water samples from an aquatic environment can be smeared onto a slide for immediate examination. Hydrogen peroxide may be added to remove the obscuring organic matter together with a small amount of hydrochloric acid (HCl) to remove any calcium carbonate from the sample. The sample is then rinsed in distilled water to remove acids completely and then can be strewn onto glass slides, dried and mounted using a thin glass plate (cover-slip) with a mounting medium with a higher refractive index than water.

Observation Techniques

Diatoms may be studied using bright-field as well as phase-contrast settings on a light microscope, the latter is better for viewing lightly silicified genera under 40 to 100 magnifications oil-immersion objectives. They may also be viewed using the scanning electron microscope (SEM).

4.4 THE LIVING CELL AND FRUSTULES

The protoplast of these unicellular organisms consists of a cytoplasmic layer that lines the interior of the frustule (silicon-cover) and surrounds a large central vacuole. The cytoplasm has a diploid nucleus and two to several chloroplast-bearing pigments for photosynthesis.

About 90% of their wall is impregnated with opaline silica. Their frustule look like a 'tablet-box' with an epitheca (larger upper valve), and a hypotheca (smaller lower valve), as shown in Fig. 4.1a. The vertical lip or rim of the epitheca is called the epicingulum, which fits over, or slightly overlaps the hypocingulum of the hypotheca. The epicingulum and hypocingulum with one or several connective bands make up the girdle. Most of them are heterovalvate (the two valves differ in shape/size).

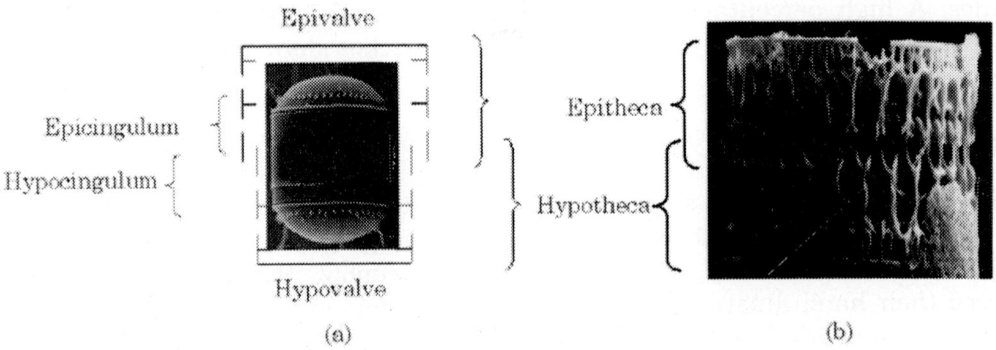

Figure 4.1 (a) Diagrammatic section and frustule of *Cosinodisia*
(b) Ddiatom filament, side view

Frustules are usually either circular (centric) or elliptical (pinnate) in valve view belonging to order Centrales and Pinnales, respectively. Nearly 10-35% of the valve has its surface perforated by tiny punctae, whose arrangement is significant for classification. The punctae may have a single hole or have thin plates of minute pores, sieve membranes. Arrangement of the punctae in lines give rise to striae, usually separated by imperforate ridges called costae.

The chain-forming species with cells linked together by siliceous structures may, in addition, have separation valves. These valves are morphologically different from the valves within the chain. Therefore, one species may have four morphologically distinct types of valves.

4.5 ECOLOGY

Most diatoms are autotrophic and form the base of the 'food-chain' in the water-bodies including oceans. As mentioned earlier they occur both as benthics and planktics in almost every aquatic environment as single cell or in colonies. They are non-motile, or capable of only limited movement along a substrate by secretion of mucilaginous material along a slit-like groove or channel called a raphe.

Pinnate diatoms dominate freshwater, soil and epiphytic niches although they can also live as benthos in marine waters. Centric diatoms, however, live as planktons in marine water, particularly at subpolar and temperate regions, but can also thrive as planktons in fresh-waters. However, normally each environment is characterized by different assemblages of species. Being autotrophic their occurrence is restricted to the photic zone (200m depending) during life. The different species have different range of salinity, temperature, pH, oxygen and other environmental parameters. In high latitudes, during summers when seasonal fluxes of these parameters change their number may rise as high as 1000 million cells per cubic meter. They are also abundant in regions of oceanic upwelling.

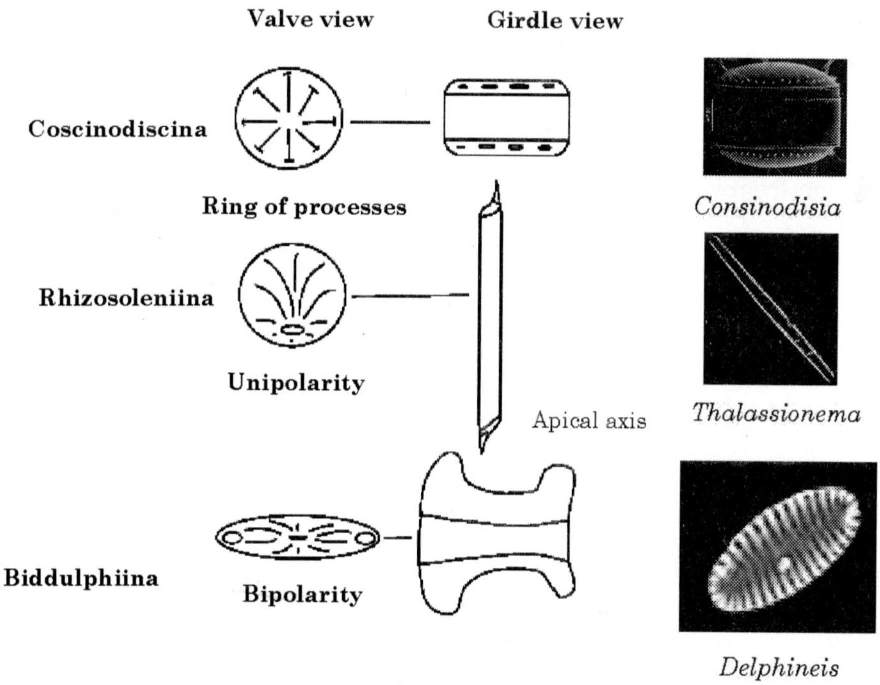

Figure 4.2 Schematic diagram of centric diatom suborders (after Hasle and Syverstsen, 1997)

Favourable conditions at their optimum level may cause accumulation of diatoms in lakes and oceans to form siliceous deposits called diatomites with up to 400 million valves per gram. Such oozes around 4-5 m thick have been found in the abyssal planes of the equatorial regions in the Indian and Pacific oceans. Despite of such records they are vulnerable to dissolution at depth or under alkaline conditions. As such their assemblages do not give a true picture. Even freshwater diatoms may be blown to the deep-sea, making the studies little difficult.

4.6 LIFE CYCLE

At the time of cell division that takes place between one and eight times per day the daughter cell takes epitheca, a valve of the parent frustule, and builds its own hypotheca within a few minutes. The rate of a reproduction is controlled by the available dissolved silica, but also because this method progressively reduces the average size of the diatom frustule in a given population there is a certain threshold at which restoration of frustule size is necessary. Auxospores are then produced, which are cells that posses a different wall structure lacking the siliceous frustule and swell to the maximum frustule size. The auxospore then forms an initial cell which forms a new frustule of maximum size within itself, with both, epitheca and hypotheca (Fig 4.3). Many neritic planktonic species alternate between a vegetative reproductive phase and a thicker walled resting cyst or statospore stage. The siliceous resting spore commonly forms after a period of active vegetative reproduction when the nutrient-level is depleted. Statospores may remain entirely within the parent cell, partially within the parent cell or be isolated from it. If nutrient-levels rise and/or length of daylight prolongs it causes the statospore to germinate and return to its normal vegatative state. So the nutrients play a key role in the alternation from vegetative stage to statospore stage.

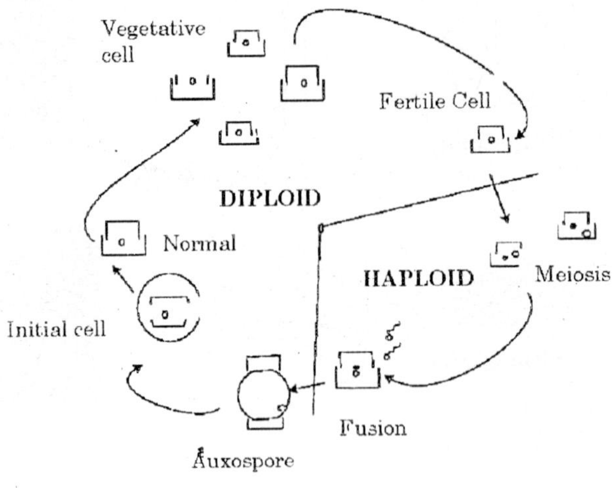

Figure 4.3 Simplified life-cycle of a common diatom

4.7 CLASSIFICATION

The valve face of their frustule is ornamented with pores (areolae), processes, spines, hyaline areas and other distinguishing features. Such skeletal features especially 'frustule-form' and 'sculpture' are used to classify and describe diatoms. Their classification was first attempted to give the group a status of division Bacillariophyta (Hustedt, 1930). Hendy (1964), however, regarded them as a class in Phylum Chrysophyta with two orders Pennales and Centrales. Simonsen (1979) and Round *et al.* (1990) also attempted a classification on diatoms. Brasier (1980) assigned diatoms to a Class Basillariophyceae of Division Chrysophyta. It has been summrised in Table 4.1

4.8 APPLICATIONS

The evolutionary history of diatoms has been utilized to allow basin wide biostratigraphic correlations. Their use in palaeogeographical interpretations for paleoceanographical purposes and a high-percentage of living genera make them valuable tools in paleoenvironmental studies. Deep-sea history researches show that they may be used in biostratigraphic zonations. They are the primary means of correlating high-latitude deep-sea sediments.

Table 4.1 Classifications of diatoms

Order	Suborder
Order Pennales or Bacillariales:  *Fragilaria* 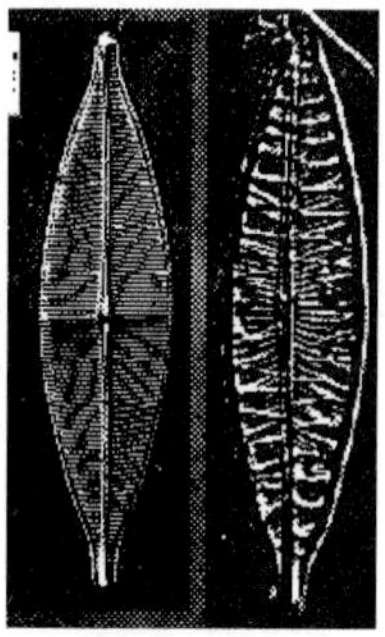 *Bacillaria*	Frustules elliptical or rectangular in valve view with bilaterally symmetrical sculpture. The latter includes a longitudinal unsilicified groove called as **raphe** down the middle of each valve face, with rows of punctuate arranged at right angles on either side (Fig. 4.3). They have valve striae arranged in relation to a line and tend to appear bilaterally symmetrical. They are divided into two sub-orders: **Suborder Fragilariida** lacks a raphe (araphid) e.g., *Fragilaria* **Suborder Bacillarida** posses a raphe e.g., *Bacillaria*
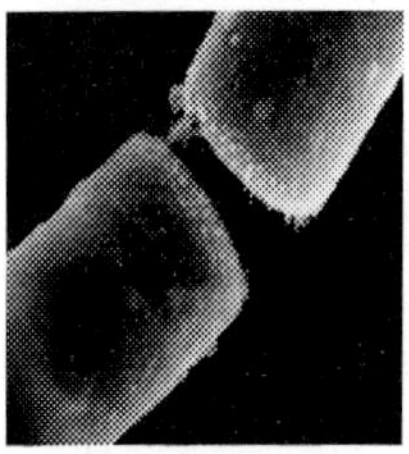 *Melosaria*  *Coscinodisia*	Frustules centric and circular, triangular or quadrate frustules in the valve view. They have valve striae arranged basically in relation to a point, an annulus or a central areola and tend to appear radially symmetrical. Being mostly planktonic and non-motile, they do not possess raphe and pseudoraphe e.g., *Melosaria* that thrives in freshwater and brackish-water habitats. Three suborders: based primarily on the shape of the cell, the polarity and arrangement of the processes.
 *Rhizosolenia*  *Biddulphinia*	**Suborder Coscinodiscina** has a marginal ring of processes and no polarity to the symmetry e.g., *Coscinodisia* **Subsorder Rhizosoleniina** has no marginal ring of processes and unipolar symmetry e.g., *Rhizosolenia* **Subsorder Biddulphiina** has no marginal ring of processes and bipolar symmetry e.g., *Biddulphinia* **Subsorder Biddulphiina** has no marginal ring of processes and bipolar symmetry e.g., *Biddulphinia*

Dissolution of frustules of the diatoms during their descent through the water column, on the sediment surface and during diagenesis may seriously alter the preserved assemblage by preferentially dissolving more lightly silicified forms. High alkalinity, burial temperatures over 50°C cause high dissolution. If the frustules sink within faecal pellets or muciligenous aggregations, rapid burial and the formation of heavily silicified resting spores tend to counteract these problems. As per an estimate only 1% to 5% living marine-assemblage is represented on the sediment surface as dead-assemblage. Still they are used to a certain extent under-utilised group in terms of biostratigraphy.

Like other unicellular organisms diatoms also show limits for tolerance to salinity, temperature and other environmental. Since a high number of fossil genera and living species, it is possible to use their ecology in deducing accurate paleonvironmental reconstructions especially in palaeolimnology.

The studies under the Deep Sea Drilling Project (DSDP) and the Ocean Drilling Program (ODP) led to development of their biostratigraphy for most of the Cenozoic particularly in the high latitude sediments, which have few/sparse or low diversity calcareous microfossils.

Diatomites consisting of 90% of frustules are mined for their economic use in filters, paints, toothpaste etc.

4.9 GEOLOGICAL DISTRIBUTION

Since the reports of the first fossil diatoms from the Jurassic could not be substantiated their appearance is considered from the Aptian-Albian stages of the Cretaceous. However, shows that naked diatoms did exist even before Cretaceous.

As with coccoliths, the earliest diatoms in the fossil record reflect a biomineralisation event rather than an evolutionary appearance. The earliest araphid (lacking a raphe) pennate diatoms appeared in the Late Cretaceous, and raphid pennates in the Middle Eocene.

The first freshwater diatoms appeared in the Palaeocene in Russia and in the Late Eocene in North America. Like Radiolaria diatoms also show a gradual progression towards less heavily silicified frustules, probably as a result of increasing competition for a limited resource (silica).

CHAPTER 5

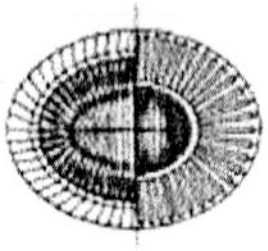

Calcareous Nannoplanktons

Contents

5.1. INTRODUCTION

Calcareous nannoplanktons (Gr. *nanno*, dwarf, ranging in diameter from as small as 0.25 to 30 μm, 1μ = 1/1000 μm) constitute a group of different morphological forms including coccoliths (3-15 μm), discoasters (5-25 μm) and nannoconids (0.20 to 30 μm). Coccoliths show clear relationships with the living coccolithophores, (having button-shaped coccolith-shields) while the other two show no clear lineages with any living organism. Since all of them are of same size and occur together in a nannoplankton assemblage, they are traditionally studied together by nannopaleontologists.

Their living species are important constituent and major source of other planktons in the marine food-chain. Because of their dual characteristics (plant and animal), diverse nature and some times unknown affinities of the morphological groups of nanno-planktons, researchers in early years could not satisfactorily classify them (Haq, 1978).

Recent researches show that the Coccolithophores are unicellular planktonic protozoa with chrysophyte-like photosynthetic pigments that differ from most of other Chrysophyta in having two flagella of equal length and a third whip-like thread, haptonema.

They occur in almost all the fine marine Tertiary sediments especially the limestones for e.g., the upper Cretaceous chalks all over the world. They are visible under a high magnification of X500 to X1000, showing great diversity and the quantitative prominence of tiny rosettes or arrays of calcite crystals. The coccoliths form a shield of 'button-like' structures (ultra microscopic in nature) around the cell, and fall apart on the ocean floor to form deep sea oozes and fossil chalk. Over 200 known species of these minute unicellular, autotrophic marine algae, known as coccolithophores are also found in the modern oceans. They belong to the Class Coccolithophyceae, Division Chrysophyta, and Kingdom Protista, based on their chrysophyte like photosynthetic pigments. But unlike other Chrysophyta they have two flagella of equal length and a third whip like organ known as 'haptonema'. The fossil coccolithophores are related groups of nannofossils (collectively known as nannoliths) that formed important constituents of marine carbonate sediments since early Jurassic times.

Their great abundance, tiny size and easy recovery from the sediments make them very significant for biostratigraphic zonation and correlation of the post-Triassic rocks.

5.2 HISTORICAL RESUMÉ

The history of their studies dates back to 1836 when the famous German biologist C. A. Ehrenberg reported very small, flat, nearly button-shaped or elliptical discs of inorganic or 'agaric mineral' or 'calcareous calcite discs' in the samples of chalk from an island of Baltic Ocean. In the subsequent years also they were regarded as 'morpholiths' or 'crystalloids' and their inorganic origin continued to be accepted for about three decades. Huxley (1858-1968) reported the presence of Ehrenberg's 'crystalloides' in deep-sea oozes recovered prior to the laying of the first trans-Atlantic telegraphic cable and referred to as coccoliths of inorganic origin.

Sorby (1861) and Wallich (1877) concluded, separately, that coccoliths were parts of larger spherical objects to which the former gave the name 'coccosphere'. Wallich also attempted to compare the coccospheres to juvenile stages of foraminifera. Their peculiar optical properties gave them a separate identity. Wallich (1863 to 1877) clarified the true nature of the coccospheres that swim freely near the surface of the tropical waters of the Indian and Atlantic oceans.

The deep-sea sediments collected during the famous H. M. S. Challenger were examined by Murray and Renard and one out of the 35 volumes published in 1891 was devoted entirely to the varieties of fauna and flora, including nannoplanktons. Accordingly the rhabdoshperes (with funnel-shaped coccoliths) and coccospheres are common in the tropical ocean, but the former quickly diminished in numbers in the higher latitudes, (rarely occurring in the surface water temperature less than 18.5°C). On the other hand, the coccospheres were encountered in the surface-water temperature, as low as 7.5°C.

In the beginning of the nineteenth century they acquired a practical importance. Aspects like their importance in the food-chain of the oceans; and biological, structural, and stratigraphical aspects received emphasis: Hans Lohmann studied the living forms from the Mediterranean Sea; L. Rabenhorst in early ninetieth-century published a classic work on the *Kryptogamen-flora;* Josef Schiller (1930) presented a comprehensive account of all the known species of extinct coccolithophores; F. Bernard (1953-74), and G. Deflandre (1939-54) published a series of papers; and E. Kamptner (1931-54) coined the term 'nanno-paleontology'.

Their importance as biostratigraphic indicators was revealed in 1954 when M. N. Bramlette and W. R. Riedel suggested their usefulness in the world-wide correlation of pelagic sequences, particularly of Mesozoic and Tertiary ages. In the late 1960's the knowledge of the fossil species and their vertical ranges reached to a significant level. Stratigraphic zonation was proposed by W. W. Hay, H. M. Bramlette and B. U. Hag that could be successfully used for worldwide correlation of Tertiary strata. Accordingly, they established practical biostratigraphic zonation of Tertiary strata. Leoblich and Tappan (1966 to 1973) published their annoted index and bibliography. Hanjo and Okada (1974) have shown that the surface water temperature is an important control-measure of distribution. It give rises to different assemblages having their adaptability to different climatic zones. Haq and Van Couvering (1977) used them for the correction of Pliocene/ Pleistocene boundary.

5.3 TECHNIQUES OF STUDY

Sample preparation techniques

There are four methods of preparation techniques concerning the calcareous nannoplanktons.

Smear slides: A small part of sample is scraped onto a glass slide and mixed thoroughly with distilled water. The slide is then dried on a hot plate and a cover slip is mounted using Canada balsam in xylene. The method is quick but some of the forms may be obscured due to the high percentage of rock-powder.

Wet sieving: A rock is mechanically crushed and boiled for half an hour to break up the small agglutinated particles. The material is then washed through a series of sieves, in order to separate out larger detritus, and then the residue containing the calcareous nannofossils is collected.

Gravity setting: This method is used to separate particles of different size and density. The samples are taken using a pipette after different settling times to obtain solutions containing different size particles. Each sample is then mounted on a glass-slide with Canada balsam and examined under a microscope. In a different way the sample may first be soaked in hydrogen peroxides and, after settling, the residue is placed in a filter-funnel with 90 to 150 μ m porosity to separate the nannofossil-size fraction.

Centrifuging: It involves centrifuging of the sample at 300 rpm for 15 seconds. The sample is then splitted into five subsamples each of which is centrifuged at different rates and for different times. This method results in the highest concentration of nannofossils, but diversity is low. Therefore, the methods of wet sieving and gravity settling are the most useful and recommendable.

Microscope techniques

The aim of microscopic techniques is to understand the structural relationship of the various elements which comprise the nannofossil. The techniques are:

Plolarizing microscope techniques: A few drops of solution containing the nannofossils are spread thinly onto a glass slide, which is allowed to dry, and a cover slip is mounted using Canada balsam in xylene. In a method suggested by Bramlette and Sullivan (1961) hydrofluoric acid is used to replace the calcite of nanno-fossils with fluorite. But it restricts the use of polarized light as fluorite is isotropic. The method also makes the nannofossils non-suitable for electron microscopy. Structural studies of nannofossils may be carried out in detail between cross nicols.

Electron microscope techniques: Such studies may be carried out using: transmission electron microscope (TEM), that use carbon replica, or scanning electron microscope (SEM). The latter is preferred as it is easy, fast and gives good-results. The preparation of material for electron microscopy, a few drops of solution containing nannofossils are spread thinly on a ground glass cover slip, which is left to dry for some time in an oven below 40°C. After ensuring the evenly distribution of nannofossils under a polarizing microscope, the cover slip is stuck on a stub of aluminium. The surface of the stub is then coated with a thin layer of gold/platinum (500 Å in thickness) to make it perform like a conductive plate.

The use of both the polarizing and electron microscope is recommended. However, for the practical purposes and for rapid results the former is commonly used as this is easy and gives better results. Even very small forms can be suitably micrographed for the study of their wall and its ultrastructure.

5.4 THE LIVING ORGANISM

General

The living coccospheres is a unicellular algae belonging to the Chrysophyceae that includes very primitive plants related to the seaweeds. They are too small to be collected in the planktonic net, and usually obtained by centrifuging water-samples from oceans.

Most of them are free swimming using a pair of flagella arising from one pole of the cell. Like other plants, they are autotrophic and use sunlight to photosynthesize the food from carbon dioxide and water, and possess chromatophores for this purpose.

A coccolithophore is generally a spherical or oval unicell of <20 μm size, equipped with two golden brown pigment spots with a prominent nucleus located between, mitochondria, the dictosome or the golgi-apparatus, and a vacuolar body, the organic surface scales, two flagella of equal length and a whip like haptonema similar to the algae of class Haptophyceae (Fig. 5.1). Their ability to secrete calcareous plates or coccoliths, distinguishes them from the other members of the Haptophyceae and all other algae.

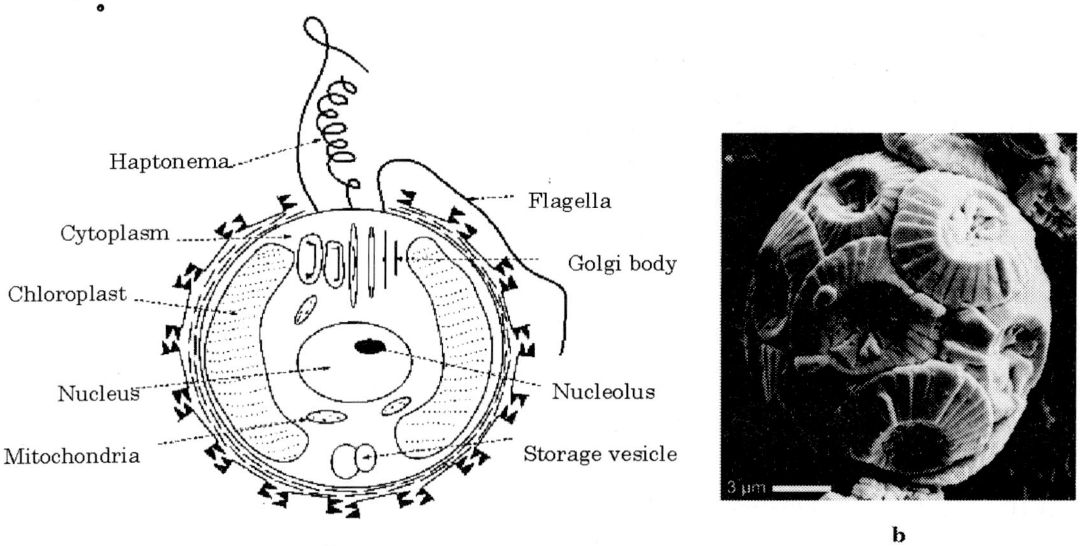

Figure 5.1 a. Diagrammatic section of a living cell of *Cycloccolithus* (3,500X), **b.** SEM picture of *Coccolithus pelagicus*

The small, calcite coccoliths are formed in vesicles within the cells in presence of light. They pushed to the periphery of the cell where they replace others. Reproduction is mostly asexual, by a simple cell division (mitosis) of the mother cell into two or more daughter cells. A few genera are characterised by alternation of generation between a 'motile' planktonic and a 'non-motile' benthic stages, with flexible skeleton with coccoliths embedded in a pliable cell membrane, in the former and cysts calcification of the membrane making it a rigid and preservable shell known as coccosphere, in the latter. They range in age form upper Triassic to Recent.

Life cycle, nutrition and growth

Two phases of life cycle has been observed in a few species of coccolithophores:

(a) Motile phase, during which they possess the flagellar apparatus, the haptonema, alternating with non-motile phase. The cell generally secretes coccoliths

(b) Non-motile phase, during which the cell either remains naked or bears coccoliths of different types than those in the non-motile phase.

Coccolithophores are photosynthesising (autotrophic) organisms. However, in some cases the phenomenon of phagotrophy (ingestion of foreign organic matter) has been reported. Several species are believed to be completely heterotrophic (utilizing both organic and inorganic substances).

In general, the growth rates of coccolithophores are relatively high, and a few species are known to multiply more than twice in a day. Most of the species exist within a short range of temperature (e.g., in *Emiliania huxleyi*) from 7° to 27°C.

Coccoliths

Process of formation

In a cross-section of live *Emiliania huxleyi* (Wilber and Watabe, 1963) the mineralization of the coccolith may be seen within the protoplasm immediately next to the reticulate body (Fig. 5.2). Between this body and the nucleus, a non-granular matrix material is differentiated that first assumes the general shape of a coccolith. After their formation the coccoliths are pushed to the surface of the cell forming an interlocking spherical envelope around and may become multilayered as more and more coccoliths are produced and pushed in the same manner. Most coccolithospheres have a definite range in the number of coccoliths around the cell, but these ranges vary considerably in different species. The shape of the envelope itself may vary from the most common spherical to near-spherical, oblong or spindle shaped.

Mineralogy

The commonest polymorph of calcium carbonate in coccoliths is hexagonal calcite. A few orthorhombic and mineralogically aragonitic forms are also known in Recent species. Wilbur and Watabe (1963) found that the two polymorphs of calcium carbonate-aragonite and vaterite, are present not only in small amounts but remain unpreservable due to their unstable nature.

Modes of construction of coccoliths

The SEM studies have revealed two basic modes of constructions in the coccolithophores:

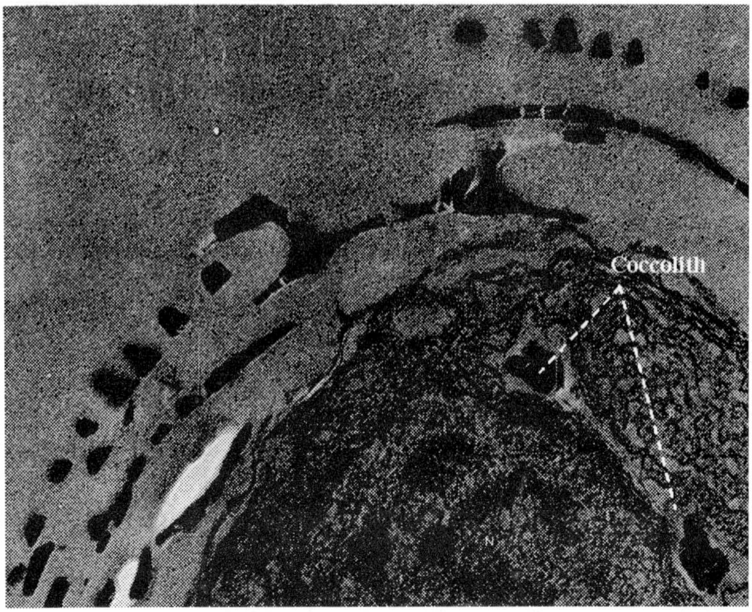

Figure 5.2 Coccolith formation in *Emiliania,* between the nucleus (N) and reticular body (R). Complete coccoliths in cross-section outside the cell at the top. x 25 000 (after Wilbur and Watabe, 1963).

Holococcoliths, *formed of similar type of crystals*: They belong to a smaller group of species that produce coccoliths of ultramicroscopic crystals, mostly, arranged in regular order, usually equidimensional and more or less maintain their original rhombohedral or hexagonal prism habits. They are disintegrated after they are shed (Fig. 5.3).

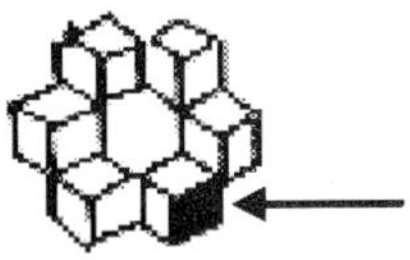

Figure 5.3 A sketch showing common type of arrangement in holococcoliths; c-axis of rhombohedra is perpendicular to the plane

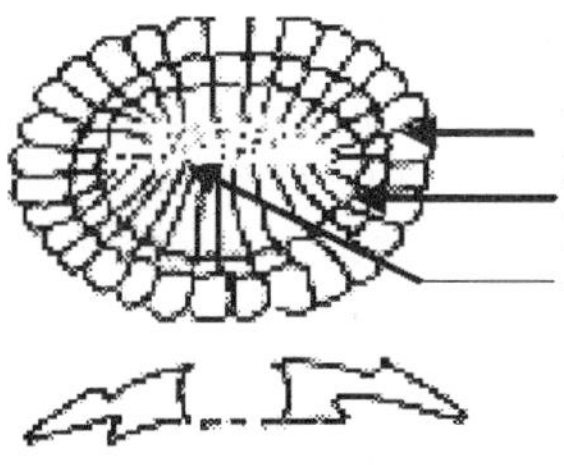

Figure 5.4 Coccolith and terminology.

Heterococcoliths, *formed of crystals of different shapes and sizes*: The majority of coccoliths of living and fossil species are made up of crystals of varied shapes and sizes in which the basic rhombohedral shape of calcite has been modified by the cell to fit into specialized morphologies. They are usually larger, rigid forms and composed of different ultramicroscopic elements such as plates, rods and grains combined together. They constitute greater part of the micro-fossil record. They also vary notably in form and construction. Their discs vary in shape from circular (shields) to elliptical and consists of radially arranged 'plates', around a 'central area' (generally hollow or porous), crossed by 'bars', filled with 'lattice' or extended in the form of a 'spine'.

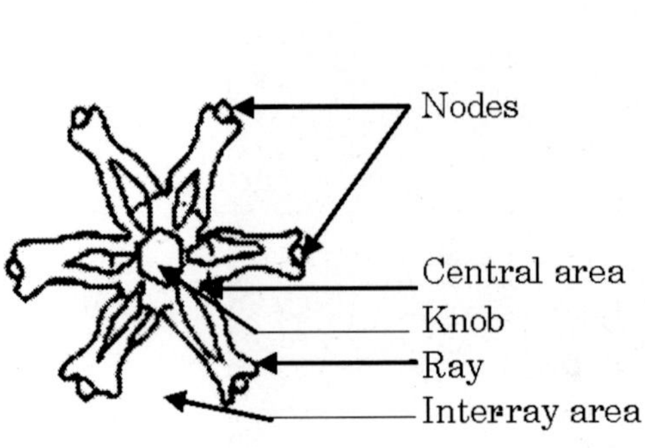

Figure 5.5 Discoaster and terminology

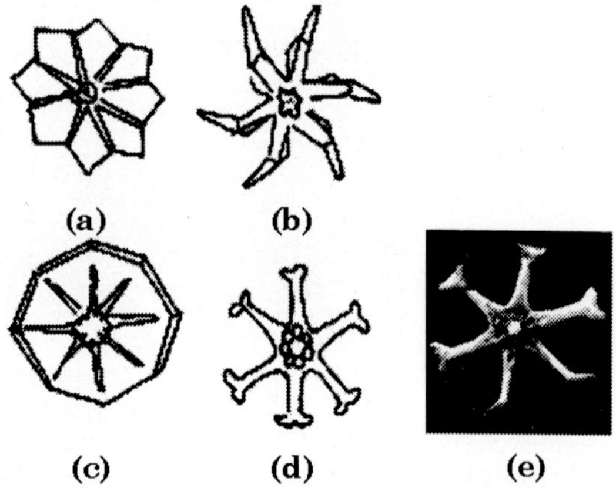

Figure 5.6. a. Sketches of *Discoaster perplexus*; **b.** *D. barbadensis*; **c.** *D. challengerii* **d.** *D. lodensis*; **e.** *D. lodensis* (SEM picture)

The coccolithophores are known to have only one type of mode of construction of coccoliths except in *Coccolithus pelagicus* which produces holococcoliths in the motile and heterococcoliths in non-motile phase of its life cycle. Completion of life cycle in coccolithophores is rare; however, a survey of literature on extant species reveals that nearly all the species that produce holococcoliths or less modified coccoliths are those that bear coccoliths only during the motile phase of the life cycle. The high flagellar action consumes most of the cell energy and the cell is unable to secrete coccoliths. Whereas, the energy consumption in non-motile phase is minimized as the flagellar action being discarded and chemical changes allowed for the secretion of coccoliths and modifications of crystal shapes and dimensions.

Discoasters and their mineralogy

Discoasters are asterisk-shaped nannofossils called asteroliths. They appeared during Tertiary and became extinct in Pliocene but have been found as biostratigraphically useful. Their taxonomy is based on the number of rays and their accessory ornamentation when seen in plan view. The number of rays is variable but the upper Tertiary species are usually have 5 to 6 rays (Fig. 5.5 and Fig. 5.6). Unlike coccoliths, they remained within the cell after calcification and not pushed to the periphery. They show very complex and varied morphologies but their ultrastructure is relatively simple as compared to coccoliths. They calcify the tabular form of calcite instead of secreting hexagonal or rhombohedral form of calcite. However, individual arms and rays are made up of single calcite crystals. They are clearly recognizable from other nannofossils by virtue of their optical character; the C-axis lie parallel to axis of symmetry making them isotropic between cross nicols.

Nannoconids

They are minute cone-shaped fossils (5-30 μm) constructed of tiny calcite wedges closely packed, in the form of a spire, the pointed end of the wedges being directed towards the centre. A canal penetrates the axis of the cone and shows great variability in width (Lord and Taylor, 1982). Colom (1956) suggested that the nannoconids are not independent organisms. Trejo (1960) gave the further evidence in which groups of nannoconids had 12 specimens arranged in a petal like structure. Biostratigraphically, nannoconids are useful in the most of the zonal schemes of the Cretaceous period.

5.5 ECOLOGY

Coccolithophores are exclusively planktonic autotrophic nannoplanktons. Their ecology is greatly influenced by the following environmental factors:

Depth and temperature

Being photosynthetic, they largely occur restricted to the photic zone of the water column (0 to 200 m). However, the maximum number is found between 100-150 m of the oceans where they remain protected from the high wave-action, on one hand and enjoy the maximum possible sunlight for their photosynthesis, on the other. Their lighter cells live near the water-surface and the heavier ones in the lower levels. The concentration decreases further rapidly at greater depths. They are distributed in open ocean, pelagic environment, near shore littoral and also in the inshore lagoonal environments. The surface water temperatures are important controls of distribution, giving rise to different assemblages adapted to subglacial, temperate, subtropical and tropical latitudes. In the tropical areas, their maximum abundance may go up to 100,000 cells per litre of sea water in the photic zone.

Salinities

The coccolithospheres thrive in littoral, lagoonal and estuarine areas where salinities are either much higher or lower than the average salinity of the open ocean (35‰). Haq (1978) demonstrated their tolerance to a wide range of salinities e.g., *Emiliania huxleyi* can tolerance-range of salinities from 45 to 16 ‰, and *Criscosphaera,* from 4 to 8‰ in the motile phase of their life cycle, and as high as 236‰ during the non-motile phase, in culture studies. Recent studies show except a few species, adopted to fresh or brackish waters the majority are marine.

5.6 CLASSIFICATION

The nannoplanktons have been claimed by both the botanists and zoologists because of their dual plant and animal nature.

Haq (1978) arranged them into three groups: (a) those that show clear relationship to the coccolithophores; (b) non-coccolithophores; and (c) common *incertae sedis* genera.

Table 5.1 shows the classification modified after Haq (1978). The 'nannoconids' (Colom, 1956 and Trejo, 1960) included in the Genera *Incertae Sedis* have been considered as a separate group (Lord and Taylor, 1982). With the exception of four groups (coccolithids, zygodiscids, braarudosphaerids and thoracosphaerids) that occur in both Mesozoic and Cenozoic sediments all other morphological groups are restructured to either Mesozoic or Cenozoic.

5.7 PALEOECOLOY OF NANNOPLANKTONS AND ITS APPLICATIONS

Most coccoliths now live in the open ocean maintaining symbiotic relation with algae. Being photosynthetic, they are mainly concentrated in the top euphotic zone. In the tropical waters they can not withstand the high surface-temperature conditions and shift to deeper levels. Many physico-chemical factors influence the ecology of nannoplanktons, as discussed. The occurrence of *Nannoconus* sp. was favoured by epicontinental and large shelf sea areas, their relative abundance decreasing with distance from the shelf and continental edge (Thierstein, 1976).

Carbonate dissolution

This hinders the preservation of calcareous nannoplankton assemblage. It begins soon after the death of an individual that sinks into water undersaturated in calcium carbonate. With increasing depth, the coccoliths dissolve/disaggregate into finely dispersed carbonate matter. Their dissolution itself depends upon hydrostatic pressure, high CO_2, low O_2, low pH, low temperature, low $CaCO_3$, precipitation by organisms, or sluggish recycling of $CaCO_3$ from the land. They sink to the bottom of deep oceans in about 100 years at the rate of 0.15m per day. It may, however, be checked/stopped if coccoliths sink within the fecal pellet of planktonic grazers. This way the sinking time up to the bottom of ocean floor reduces to only a few weeks (Hanjo, 1976) and also protects them from dissolution by decreasing their residence time in deeper undersaturated waters. In deeper waters coccolith isocline, occurs between 3,000 to 4,000 m depth (Roth and Berger, 1975). After which the diversity of the assemblage is decreases. Below 5,000 m, approximately at the level of calcium carbonate compensation depth (CCCD) that separates the predominantly calcareous sediments from the carbonate-poor ones, only the most resistant species are preserved.

Temperature

The abundance of coccoliths is greatest in subtropical and tropical regions. In these areas their average weight may be 26% of the total weight of the sediments. They were great rock-builders constituting > 21% of Cretaceous chalks and > 90% of the Tertiary chalks. They are fewer in sediments of subglacial waters (about 1%) where both productivity and preservation conditions are unfavorable.

Table 5.1 Key to major morphological groups of calcareous nannoplanktons

A. COCCOLITHOPORES AND RELATED GROUPS

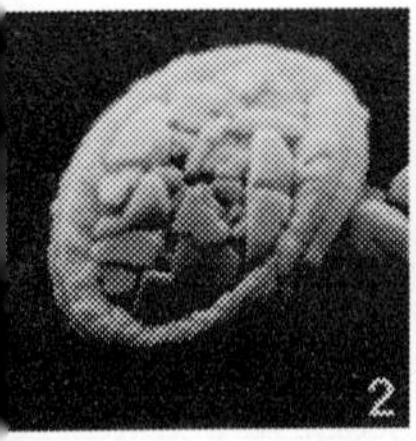

Arkhangelskiella

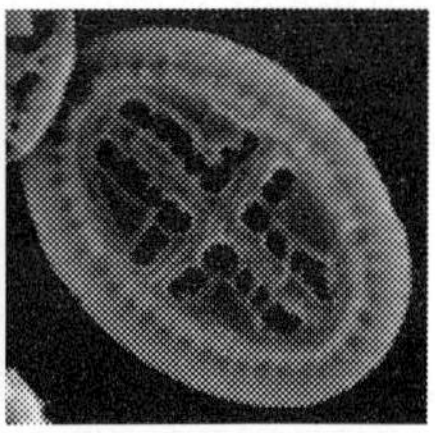

Broinsonia

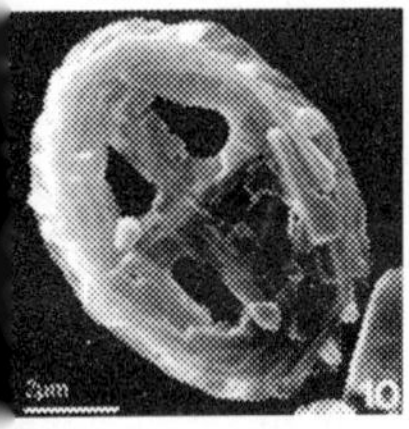

Gartnerago

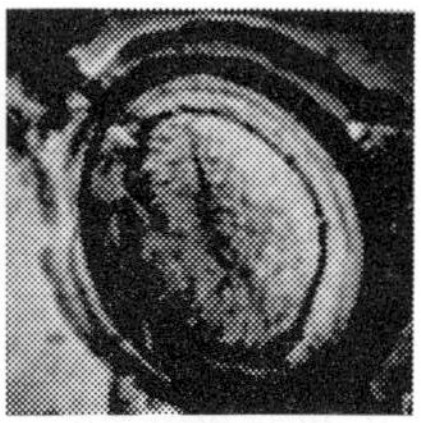

Kamptnerius

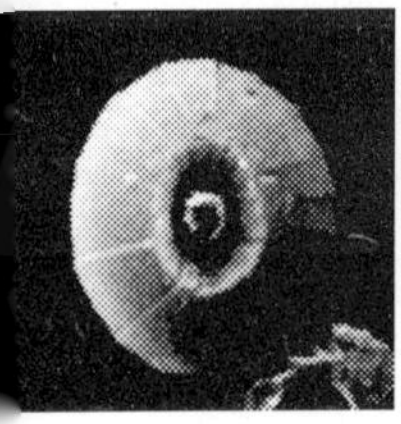

Biscutum

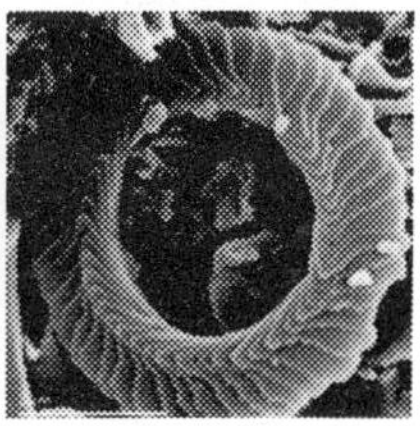

Cyclococcolithus

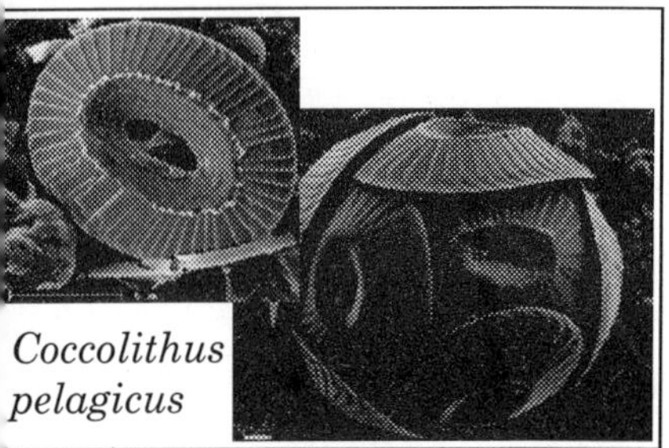

Coccolithus pelagicus

1. **Arkhangelskiellids** (Family: 1)

   Family: 1 Arkhangelskiellaceae (Mesozoic)

   Elliptical coccoliths formed of 2-4 cycles of jointed elements. e.g., *Arkhangelskiella*, has a three layered rim in proximal view; *Broinsonia*, has an addition inner rim cycle in distal view and distinctive perforations; *Gartnerago* has a multi-tiered rim in proximal view; and *Kamptnerius*, has an asymmetrical outer rim.

2. **Coccolithids** (Families: 3)

   Two shielded coccoliths, each consisting of one or more cycles of crystal elements connected at their inner margins,

   Family Coccolithaceae (Mesozoic and Cenozoic).The distinctive characters of these genera are: shape of coccolith and central area, and number of cycles in two shields. *Biscutum*, consists of two single cycled elliptical shields. Other examples are: *Coccolithus, Cyclococcolithus*

   Family Prinsiaceae (Cenozoic). Genera distinguished by structure of the rim and central area e.g., *Prinsius,* Has a multicycled distal shield but a single non-distinct central opening; *Emiliania*, has a distinct rim of I-shaped elements and a central grid.

   Family **Helicoshaeraceae** e.g., *Helicosphaera*

3. **Podorhabdids** (Family: 1). Genera are distinguished by characteristics of the central area.

   Family Podorhabdaceae (Cenozoic)

   Two shields with a single cycle of radial elements on the proximal shield e.g., *Podorhabdus,* has four radial bars supporting a central process.

4. **Pontosphaerids** (Family: 1). Genera are differentiated by the wall and central area characteristics.

   Family Pontosphaeraceae (Cenozoic)

   Elliptical coccoliths, with two closely appressed, *Pontoshaera,* thin-walled, multiple pores in central area.

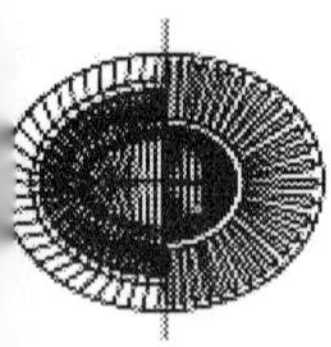

Prinsius

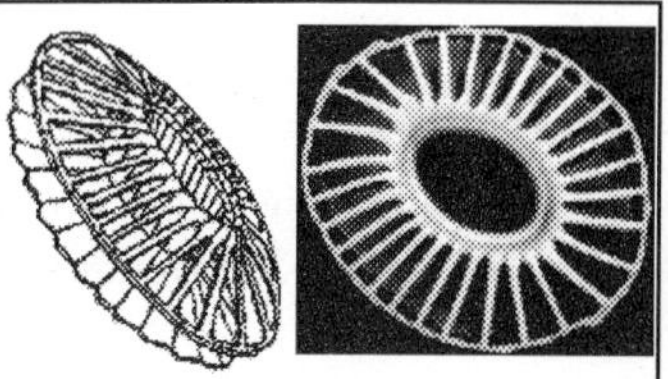

Emiliania

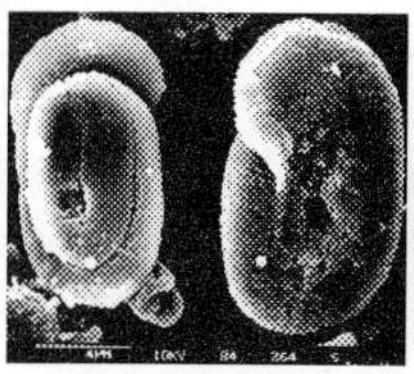

Helicosphaera

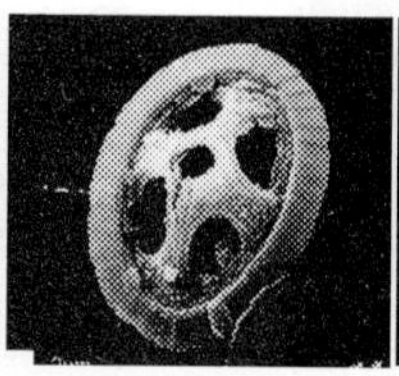

Podorhabdus

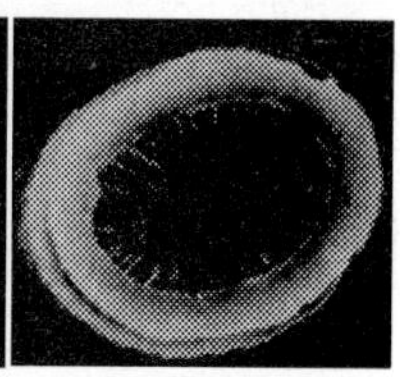

Pontoshphaera

Blackites

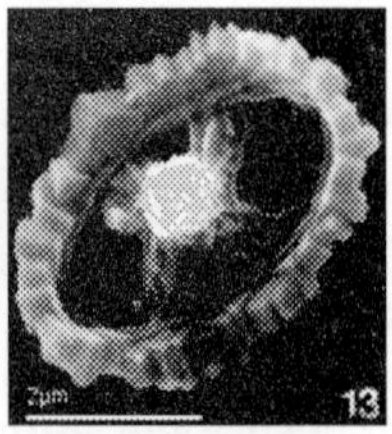

Stephanolithion

Rhabdosphaera

5. **Rhabdosphaerids** (Family: 1)

   Family Rhabdosphaeraceae (Cenozoic)

   Elliptical to circular coccoliths, with one or two simple shields e.g., *Blackites, Rhabdosphaera.*

6. **Stephanolithids** (Family: 1)

   Family Stephanolithionaceae (Mesozoic)

   Coccoliths of varied geometrical shapes, with a distal cycle of crystal elements forming a low or high cone or a cylinder and a proximal cycle of brick like crystals (*Stephanolithion*)

7. **Syracosphaerids** (Family: 1)

   Family Syracosphaeraceae (Cenozoic)

   Simple to complex wall with a characteristic central area, crystal laths extending up to the centre e.g., *Syrosphaera*.

8. **Zygodiscids (**Families: 2)

   Elliptical coccoliths with marginal wall of one or two cycles of imbricated elements and a thin inner rim,

   Family Eiffellithaceae (Mesozoic) e.g., *Effellithus*

   Family Zygodiscaceae (Cenozoic) e.g., *Zygodiscus.*

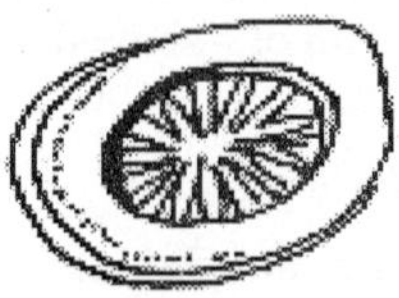

Syrosphaera

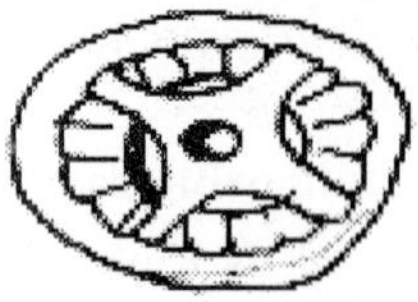

Effellithus

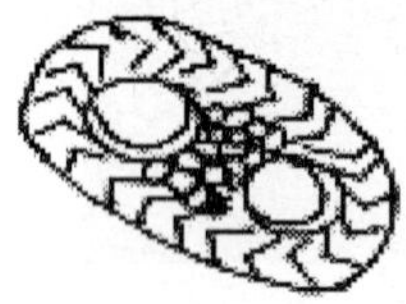

Zygodiscus

B. NON-COCCOLITHOPHORE

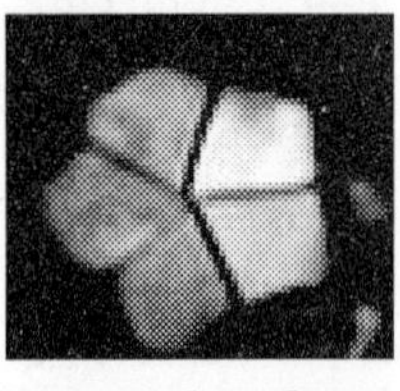

Braarudosphaera

Pemma

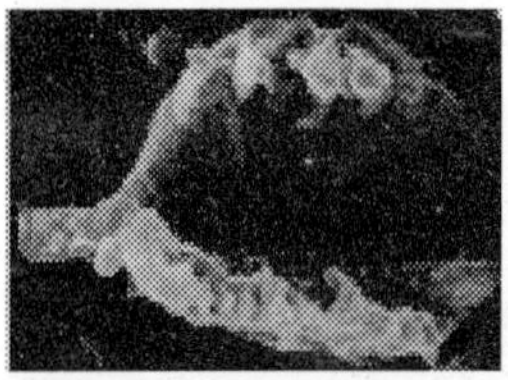

Ceratolithus

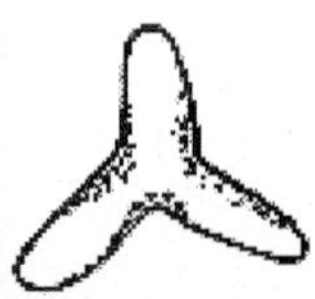

Tribachiatus

1. **Braarudosphaerids** (Family: 1)

   Family Braarudosphaeraceae (Mesozoic and Cenozoic)

   They are pentaliths (composed of 5 crystal units)

   e.g., *Braarudosphaera, Pemma.*

2. **Ceratolithids** (Family: 1)

   Family Ceratolithaceae (age: late Cenozoic)

   Horse-shoe shaped nannoliths, e.g., *Ceratolithus.*

3. **Discoasterids** (Family: 1)

   Family Discoasteraceae (Cenozoic)

   Disc- or star-shaped , with or without stems on one side of the asterolith e.g., *Discoaster, Tribachiatus.*

Non-coccolithophore nannoliths contd.

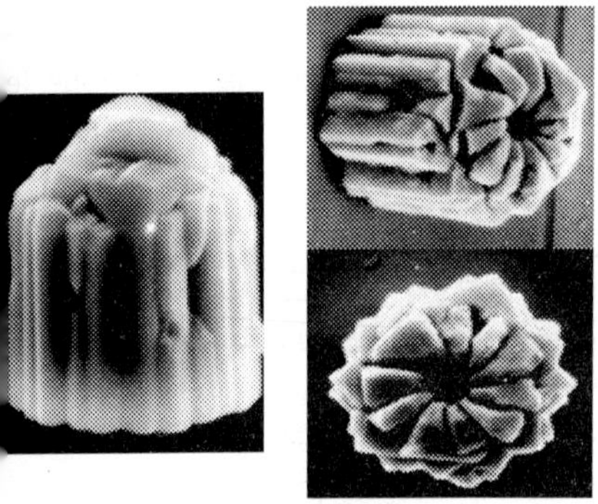

Fasciculithus

4. **Fasciculithids** (Family: 1)

   Family Fasciculithaceae (Cenozoic)

   Subcylindrical bundles of wedge-shaped crystals whose thin edges meet along the centre e.g., *Fasciculithus.*

5. **Heliolithids** (Family: 1)

   Family Heliolithaceae (Cenozoic)

   Two abutting unequal shields lacking a connecting tube and one shield with subradial sutures e.g., *Heliolithus.*

Lithastrinus

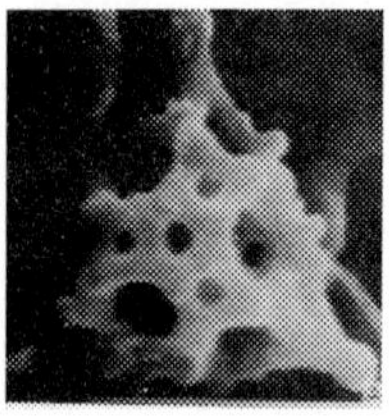

Lithostromation

6. **Lithastrinids** (Family: 1)

   Family Lithastrinaceae (Mesozoic)

   Two or more cycles of five or more imbricated crystal elements. Elements of one cycle may extend outward to form an asterolith-like nannolith e.g., *Lithastrinus.*

7. **Lithostromationids** (Family: 1)

   Family Lithostromationaceae (Cenozoic)

   Thick triangular tests with highly ornamented surface e.g., *Lithostromation.*

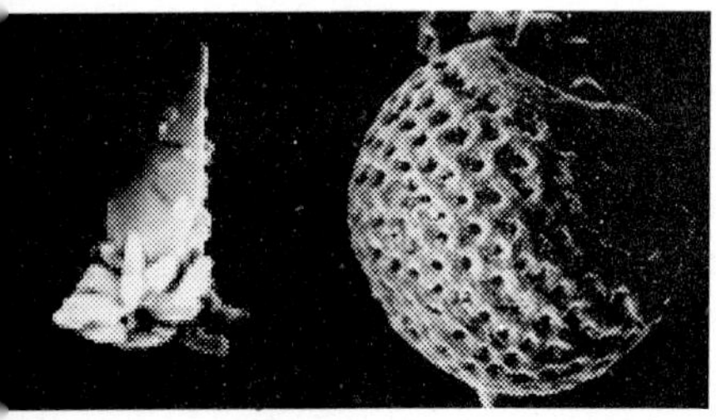

Sphenolithus *Thoracosphaera*

8. **Sphenolithids** (Family: 1)

   Family Sphenolithaceae (Cenozoic)

   They have a rounded or polygonal base, surmounted by an elongated rib-cone with one or more spines radiating from the centre e.g., *Shpenolithus.*

9. **Thoracosphaerids**

   Family Thoracosphaeraceae (Mesozoic and Cenozoic)

   Spherical shells composed of a mosaic of interlocking crystal units. They may be with or without an opening which may have lid e.g., *Thoracosphaera*

C. NANNOCONIDS

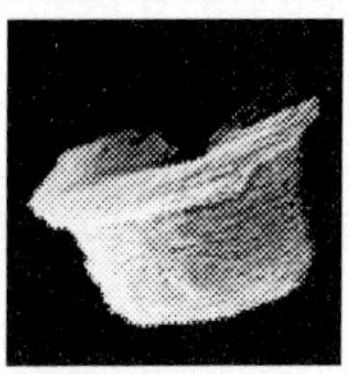

Nannoconus

Genus *Nannoconus*: Cone shaped nannoliths, composed of minute wedge-shaped crystal units arranged radially around and perpendicular to the long axis which is occupied by a canal open at both ends of the test, (Mesozoic).

Table 5.1 contd.

D. GENERA *INCERTAE SEDIS*

Isthamolithus

Microrhabdulus

Micula

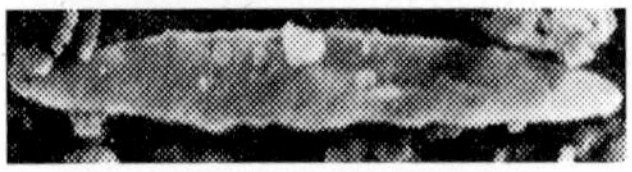

Triquetrorhabdulus

1. **Genus** *Isthamolithus*: Elongate parallelograms with wide walls and two transverse septa dividing the unit into 3 windows (Cenozoic).

2. **Genus** *Microrhabdulus*: Long cylindrical rod-like nannoliths, either truncate or tapering at ends and showing checkered extinction patterns under cross-polarized light. They may be the remnants of other unknown microfossils (Mesozoic).

3. **Genus** *Micula*: Rectangular to subcubical with concave sides and constructed of two or more units of different calcite orientations (Mesozoic).

4. **Genus** *Triquetrorhabdulus*: Rod-like nannoliths with three edges 120° apart and pointed or rounded at ends (Cenozoic).

Diagenesis

Diagenesis can also alter the nature of fossil assemblages. Recrystallization can result in heavy calcite overgrowth on crystal faces. Sometime it alters the morphology of nannoliths beyond recognition, especially in discoasters (Wise and Kelts, 1972). Thierstein (1974) found a number of Cretaceous taxa that spanned equal stratigraphic intervals to be conspecific, distinguished from each other by diagenetically induced features.

On the basis of the ecological factors the applications of coccoliths and other nannofossils may be summarised as follows:

a. Nannofossils are considered as one of the excellent groups of marine microfossils by virtue of their short stratigraphic ranges and have been used for the biostratigraphic subdivisions of Mesozoic and Cenozoic rocks.

b. Being planktonic and very small their quick dispersal has enhanced their value not only as index fossils but also in biostratigraphic zonation at very high resolution.

c. As the base member of the marine food-pyramid they are important in the oceanographic studies especially in the latitudinal zonation.

d. Their high abundance both in coastal and oceanic regions from north to south-pole makes their ecological studies important. Their seasonal-upwelling is of particular importance as it supports other unicellular organisms like foraminifera, which themselves are important in so many oceanographic studies.

e. Their studies have revealed that the last glacial maximum (c. 18,000 BP) witnessed shift of water-masses and the standing-crops by at least 15° towards south. Coccoliths-based study (Haq, 1980) also revealed a similar shift in the water-masses with their constituents in the Miocene times. They also proxy for the glacial and interglacial periods of the Pliestocene times fairly well.

f. Ratio of two of their warm and cool water species namely, *Diascoaster* and *Chiasmolithus* has been found to be a good indicator of change in temperature in the late Paleozoic time (Bukry, 1973, 1975).

g. They have also been utilized to some extent in the paleotemperature studies based on the oxygen isotopes separated from their coccoliths and have been matched with the value obtained from the contemporary planktonic foraminiferal species with remarkable similarities.

h. Being very small their isotopic studies for other elements are also useful to understand the chemistry of surface, column and bottom waters of the present and the geological past.

5.8 GEOLOGICAL DISTRIBUTION AND BIOSTRATIGRAPHY

The earliest stratigraphic record of coccoliths in the Palaeozoic is doubtful or less diversified. However, *Annulithus arkelli,* found in upper most Triassic (219 Ma) may be considered as the first record of nannofossils. The early lower Jurassic witnessed their diversification which parallels the radiation of the some dianoflagellate cysts and both may be related to oceanographic changes connected with the opening of the Atlantic Ocean at this time. These conditions led to the deposition of chalk over vast areas of the continental platforms.

Coccoliths with placolith structure first appeared in lower most Jurassic (206 Ma yr.) age. Their shields have elements which lies side-by-side and lack imbrications. The genera include *Biscutum, Podorhabdus, Striatomarginis* and *Discorhabdus.*

The first significant increase in the diversity of nannofossils is found in the early part of late Jurassic which reached to a maximum in the latest Cretaceous (Masstrichtian). The final morphological group of coccoliths to evolve in the Jurassic was the Stephanolitionaceae, with the species *Stephanolithion octum* of Bathonian age (175 Ma yr.). At the end of Masstrichtian, they became nearly extinct but for a few species which survived into Danian (65 Ma yr., the earliest Tertiary). The vast majority of coccolithophores became extinct at the K-T boundary, many of their habitats being filled by the diatoms during the early Cenozoics.

A number of new taxa evolved rapidly during the late Paleocene and early Eocene. Their maximum diversity is noticed in the early Eocene. Reason of such a high diversity may be attributed to the appearance of discoasters in the late Paleocene and their rapid diversification in the early and middle Eocene. This may have been due to climatic cooling and regression.

A relatively sharp diversification during the early and late Oligocene, followed by a second episode of increase in their abundance has been found in the middle Miocene after which a trend towards decreased diversity continued until Pleistocene.

Certain of the placolith-bearing coccolithophores, however, thrived in the cooler waters of the Quaternary era.

CHAPTER 6

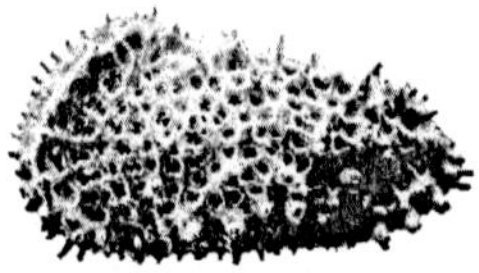

Ostracoda

Contents

6.1 INTRODUCTION

Ostracoda is a group of small bivalved crustaceans with two chitinous or calcareous valves. Because of their external similarities with lamellibranchs they are informally called 'seed shrimps'. They make the most diverse group of crustaceans having some 33000 living and fossil species.

Being crustacean, they have jointed appendages. There is no true segmentation in the soft body as often found in the arthropods but, it may be divided into two main portions: the head or cephalon, and the thorax.

Hinged above the dorsal region of the body, their valves are collectively known as carapace. Valves, unlike Lamellibranchia their valves are without growth lines and form an integral part of chitinous covering of epidermis. Their most conspicuous feature is that the carapace is shed many a times till the animals attain their adulthood. The carapaces, generally oval, bean-shaped or kidney-shaped range in size from 0.5 to 30 mm in length (average: 0.5 to 3 mm). The largest known genus is *Gigantocypris.* The left valve of the carapace is slightly larger than the right one and overlaps the latter.

They belong to Class Ostracoda, Phylum Crustacea and Superphylum Arthropoda.

They appeared in upper Cambrian as the marine and benthic (?) animals and remained so till Silurian times when they adopted to brackish (less saline) and pelagic (with in the water column) environments. Later, the brackish forms adopted to freshwater or in moist/damp soils, organic layers of the soils and hot-springs.

Ostracoda show a long and well documented fossil record since Ordovician. They are widely used in biostratigraphy, in determining paleoenvironment of deposition, and paleoclimate of the region.

Their abundance in non-marine sediments and usefulness in the biozonation of marine strata, especially for 'petroleum exploration' make them second in importance only to the foraminifera, as micro-fossils and, to none as indicators of 'ancient shore-lines', 'salinities' and 'relative sea-floor depths'.

6.2 THE SOFT BODY

Like other arthropods their visceral parts are also covered by a rigid, jointed exoskeleton of chitin. They have 7 or 8 pairs of jointed appendages or limbs (which bear fine chitin hairs or bristles or setae), attached on the ventral side (Fig. 6.1). The head is large and bears a centrally placed mouth (at ventral side). The benthic forms have a typical dorsal eye whereas; some of the myodocopids have two lateral eyes. The deep-water forms, however, are totally blind. The cephalon and thorax are indistinguishable and, therefore, called Cephalo-thorax. The body has three intimately connected features: forehead, upper-lip and hypostome, each provided with a pair of appendages. (Fig. 6.2 and 6.3). Of the seven, four of the pairs of appendages are borne by the cephalon. The first, known as antennules, consists of long and tapering appendages, mainly used for feeding. In walking and swimming, however, the second pair, known as the antennae (singular: antenna) is used. The other two pairs, mandibles and maxillae lie behind the mouth. The latter also whip-up water to provide filter-feeding currents, and respiration.

The thorax contains the stomach, intestine, genital organs etc. Three pairs of appendages, known as thoracic appendages are attached to the thorax. A pair of leg like appendages known as furcae (singular: furca), joins the thorax at its posterior side.

Their respiratory and circulatory systems are underdeveloped. They do not have large blood vessels and heart except in large planktonic myodocopids.

The muscles are used for closing and keeping the valves closed, are known adductors muscles. They are generally situated at the mid point of the valves. After the death of animals they leave scars. The numbers, shapes, sizes, and configuration of their scars make an important basis to classify them, especially the archaeocopids, leperditicopids and paleocopids.

6.3 THE CARAPACE

Morphology

The ostracoda carapaces may be oval, kidney-shaped or bean-shaped with a hinge and dorsal margin and vary in length from 0.4 to 3 mm. For the micropaleontologists their carapace is the only material from where the information regarding the anatomy, ontology, phylogeny, ecology etc., of fossilized or extinct representatives can be obtained.

The left and right valves collectively constitute a carapace. The carapace is formed by two lateral folds of epidermis, the duplicatures, which originate in the head region and extend forwards, backwards and downwards to enclose the body and limbs. Each valve is a fold of skin, as it is attached to the chitinous covering of epidermis. The two valves are hinged together at the dorsal margin, by a flexible chitinous layer known as ligament. Unlike lamellibranches, the growth in ostracoda is achieved in stages and the carapace is shed at each stage of growth. The process is known as moulting and the stages as instars. Generally, carapace is shed 8 times before adulthood (Fig. 6.1).

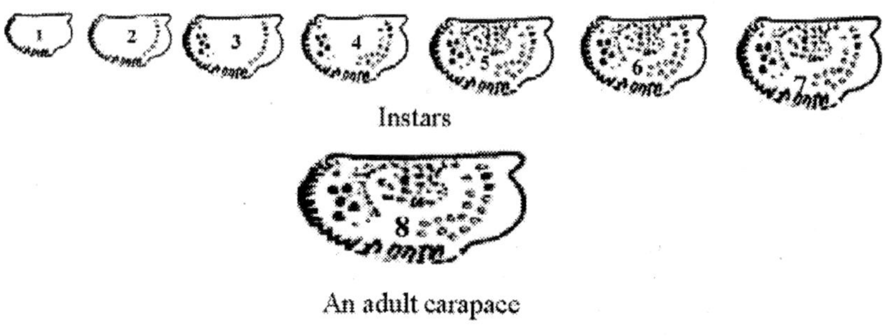

Figure 6.1 The successive instars showing molting-stages in Ostracoda

The duplicatures have an outer- and inner lamella. The former may either be fused or free from the latter and can have calcified and uncalcified parts. Each of them consists of three separate layers: outer chitinous, central calcareous, and inner calcified. In the outer lamella a space is left between the two lamellae, called vestibulum that is actually an extension of the body that in some taxa can house digestive and reproductive organs. Ridges on the duplicature in the ventral side are known as selvages (blade-like extensions of the two outer lamellae). The inner most line of the contact between the fused lamellae is called line of concrescence and the area between this and the outer margin is called marginal zone (Fig. 6.4). In podocopids the valves are formed by the secretion of calcite by the epidermis of the outer lamella and the peripheral part of the inner lamella. In the living animal the calcareous valves are thus enclosed by the chitinous epiculitcle.

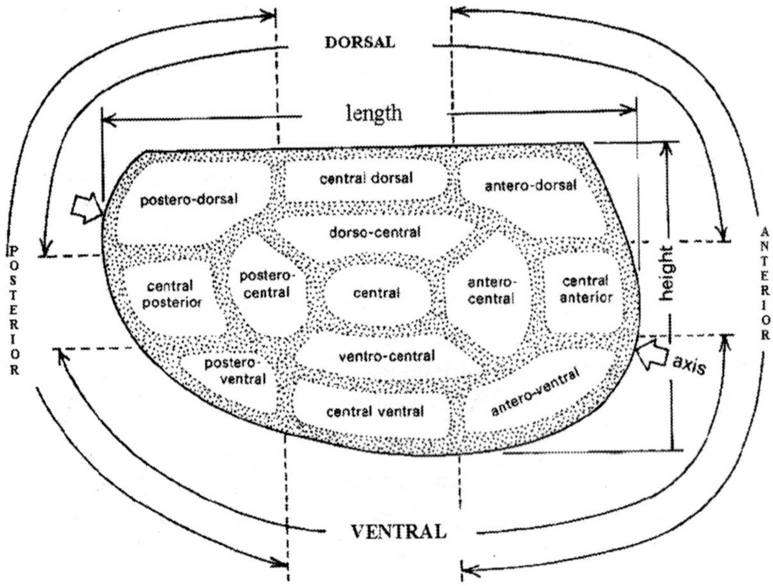

Figure 6.2. Orientation and nomenclature of the right valve of carapace (modified after Kesling, 1951).

A different terminology has been used for myodocopan, in which the likely homoloque of the calcified inner lamella is infold. So in such cases the carapace should comprise of the outer lamella and the infold, the uncalcified inner lamella and the body cuticle are the vestment. The valves of many myodocopids are not or weakly calcified, as fossils are commonly secondarily calcified. The calcification is weak in juvenile or freshwater podocopids, many myodocopids and archaeocopids; as such they left a meagre record as fossils.

In podocopids the valve are unequal. The larger one overlaps the smaller along all or the part of the margin. However, in the myodocopan such overlap is inconspicuous. The external surfaces of the inner and outer lamellae are covered with a thin layer of cuticle, which is continues across the dorsal margin and connects the valves of the ligament.

Their sensory bristles (sensilla) penetrate the outer lamella through numerous canal-like structures running across the marginal zone from the inner space to the outer world. Generally simple, except in some cytherids they are widely branched. The lateral surface of the outer lamella is pierced by normal pores or lateral pores. Those sensilla that penetrate the marginal zone are called as marginal pore canals whilst those traversing the rest of the shell are termed as normal pore canals. Their form that is, branched or unbranched; besides their arrangement is used in taxonomy. Sieve pores also with sensilla in some podocopans are important in taxonomy. However, there may be changes in their shapes, which could be linked to the salinity. Clear eye-spots and raised eye tubercles may also develop adjacent to the eyes, especially in shallow water forms.

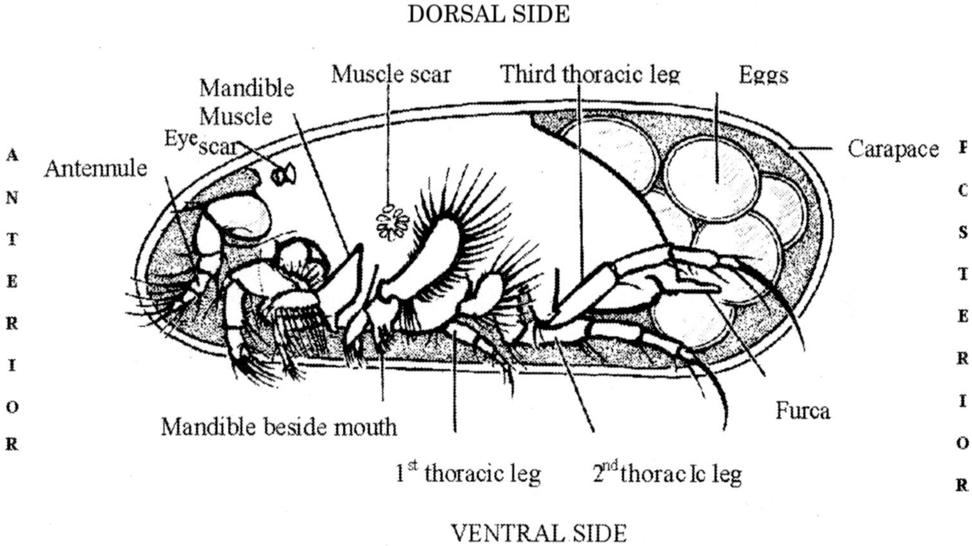

Figure 6.3 Living Ostracoda (female) with left valve removed (after Brasier, 1980)

Outer lamella

It is generally calcified throughout its whole length, except in a few myodocopids. In the extinct archaeocopids, the outer lamella appears to be mainly chitinous and secondarily calcified, or phosphatised after their death. Therefore, the juvenile and freshwater podocopids, marine myodocopids and, archaeocopids do not commonly occur as fossils (Bate and East, 1972).

Inner lamella

The calcified part of the inner lamella is secreted by the similar epidermal cells. Such calcification is weaker in juvenile (immature) stages and also in freshwater species. In most of the planktonic myodocopids, the calcification is weak and amorphous. The margins of thin chitinous inner lamella are calcified into thicker 'duplicature' in adult podocopids and myodocopids. This is secreted to strengthen the interlocking between valves. The free margin of the duplicature is called flange. The inward projection of the duplicature, which seals the two valves in the closed position, is called selvage (Fig. 6.5 and 6.6). The structure, and the degree of development of inner lamella in juvenile forms is hardly noticeable.

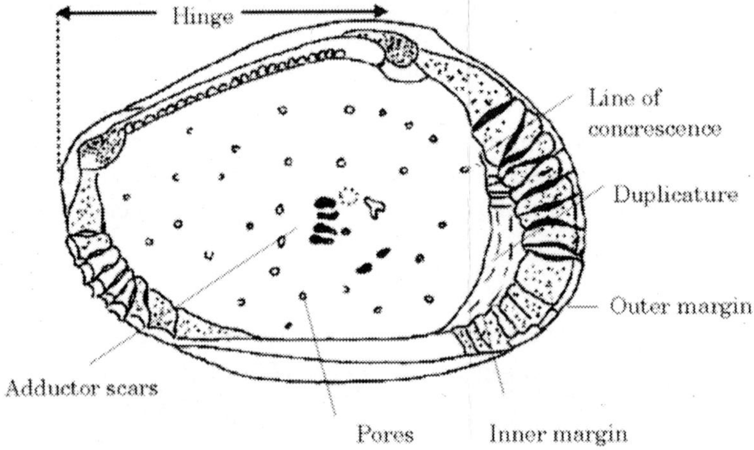

Figure 6.4 Inside view of a podocopid right-valve (after Van Morkhoven, 1962 and Kesling, 1951)

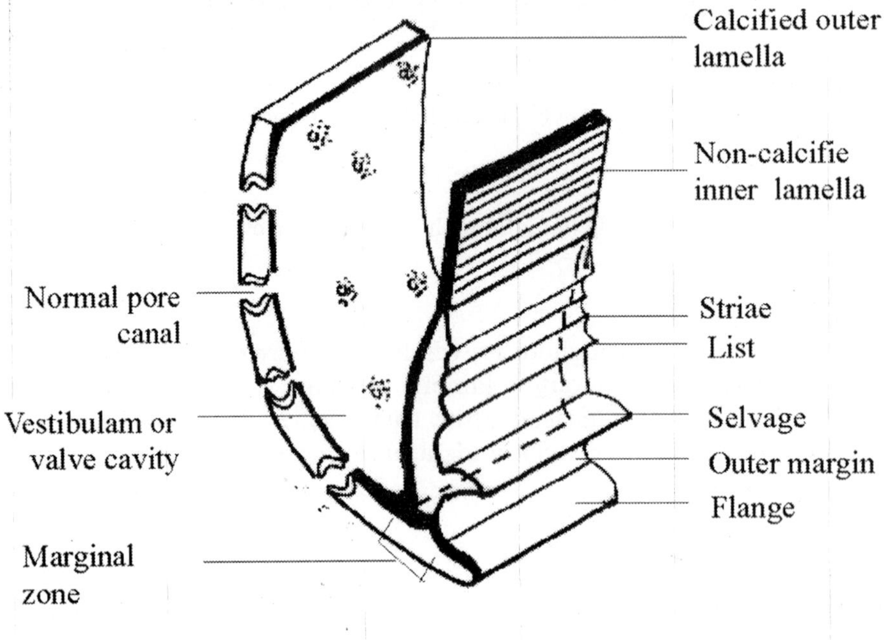

Figure 6.5 Diagrammatic-section of a part of a podocoid-valve (after Kesling, 1951)

Fig. 6.7 shows that the 'outline' of valve is the profile of the carapace which includes all the peripheral elements such as alae, caudal-processes or projections, marginal denticulations, hinge-ears, brood pouches etc. The outer margin of a valve, however, is actually the outer periphery of the valve and may be appreciated in the inside view. Several features are observed at the margins of the outline of carapace or single valve.

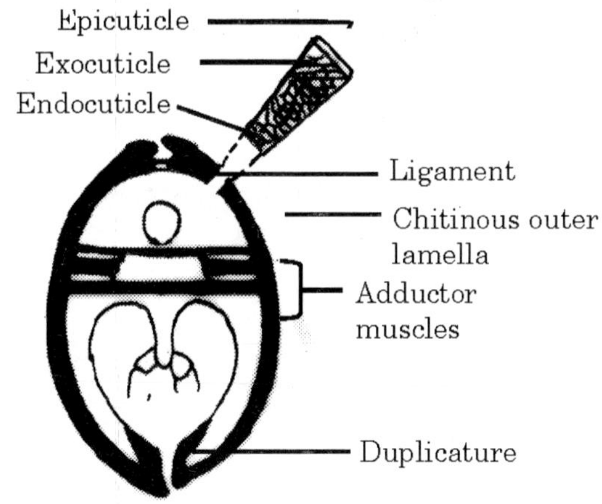

Figure 6.6 Schematic cross-section of a living Ostracoda (after Brasier, 2008)

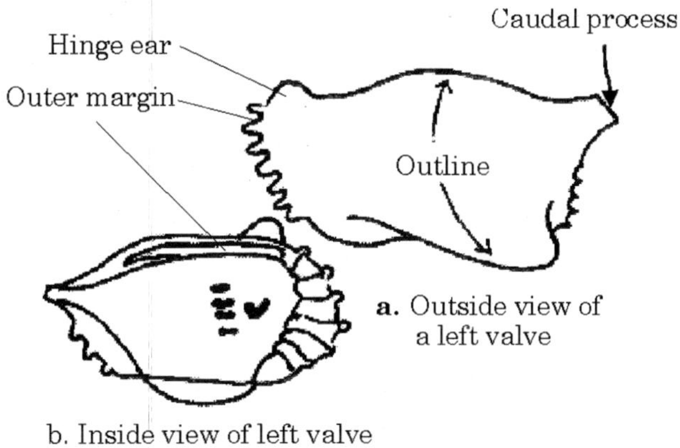

Figure 6.7 Difference in outline and outer margin

Alae or lateral prolongation The weight distribution projections developed to gain stability in the flat ventral-surfaced benthic ostracodes e.g., *Cytheropteron* (Fig. 6.8b).

Caudal processes A nozzle like projection, drawn out from the posterior end in some species. A hole always remains present at the tip of it to help in the better circulation of water (Fig. 6.7 and 6.8 c).

Posterior spines Two or more calcareous spines found at the postero-ventral angulations of some species e.g., *Cyprideis* and *Paracypreideis.*

Hinge ears Extension of the outer dorsal margin, found in the larger valves of some species.

Eye tubercle or eye spot A raised and extraordinarily large, upward pointing part of valve, situated below the antereodorsal hinge, found in some species e.g., *Indiana.*

Brood pouches or egg pouches Pronounced ventral lobes found on the posterior end of female or hetromorph form of ostracoda e.g., *Beyrichia.*

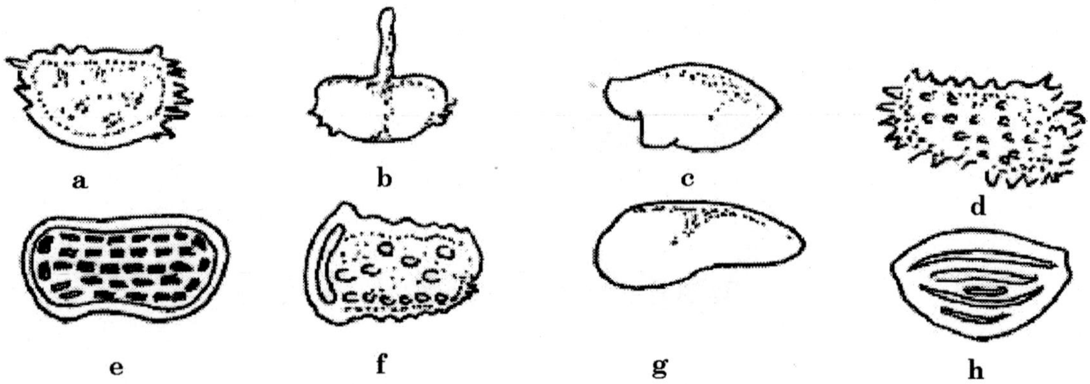

Figure 6.8 Common ornamentations : **a**. marginal denticulations (***Cypridea***); **b**. spinose (***Aechmina***); **c**. alae (***Cytheropteron***); **d**. spinose; **e**. reticulate; **f**. nodose; **g**. smooth surfaced (***Darwinula***); costae (***Richteria***)

Antero-ventral beak-rostrum- A marked abrupt indentation on each valve on the antero-ventral margin of some of the extinct cyprids (family Cypricidae).

Dorsal spine- A long tapering spine extended from the central dorsal margin in some species of suborder Beyrichicopina e.g., *Aechmina.*

Frills or keels- Thickened, calcareous border on ventral margin, to provide more stability to the carapace (or a form of weight distribution feature) e.g., *Beyrichia.*

Sulci and lobes- Depressed and raised portions, generally near dorsal margin in some species e.g., *Kloedenella* (Fig. 6.9).

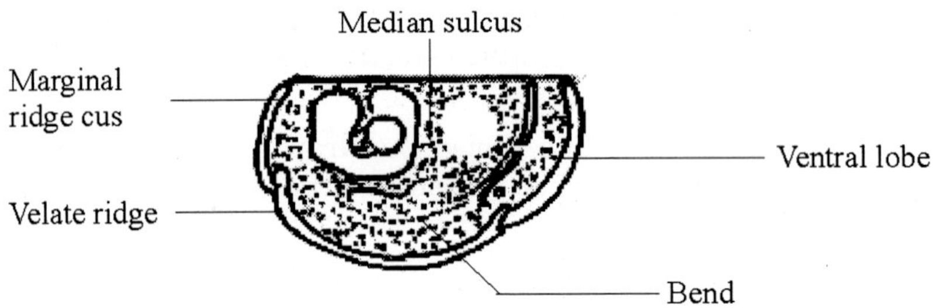

Figure 6.9. Morphology of a palaeocopid valve (lateral view)

Hinge and articulation

The hinge, dorsal by position, constitutes the line and a zone (hinge zone) along it. It may be straight (*Beyrichia*) or convex (*Healdia)*, and may not occupy the whole length of the dorsal margin of the carapace. The hinge zone may have some structures to prevent any translational and rotational movement between the valves and called articulation. Myodocopids and cladocopids lack articulation, but rather have an elastic-ligament, along the hinge lines to keep the valves firmly attached. In others, ridges, grooves, teeth and sockets of simple or crenulated outlines are found in the hinge zone (Fig. 6.10).

The type of hinge articulation in Ostracoda is important for their classification. Van Morkhoven (1962) has classified the hinges and their elements into:

(a) Simple hinge or adont hinge showing a single element in the form of a groove on the dorsal margin of the larger and the corresponding socket in the smaller valve;

(b) Compound hinge with three elements: anterior, median and posterior; and

(c) Simple hinge with ridges and grooves.

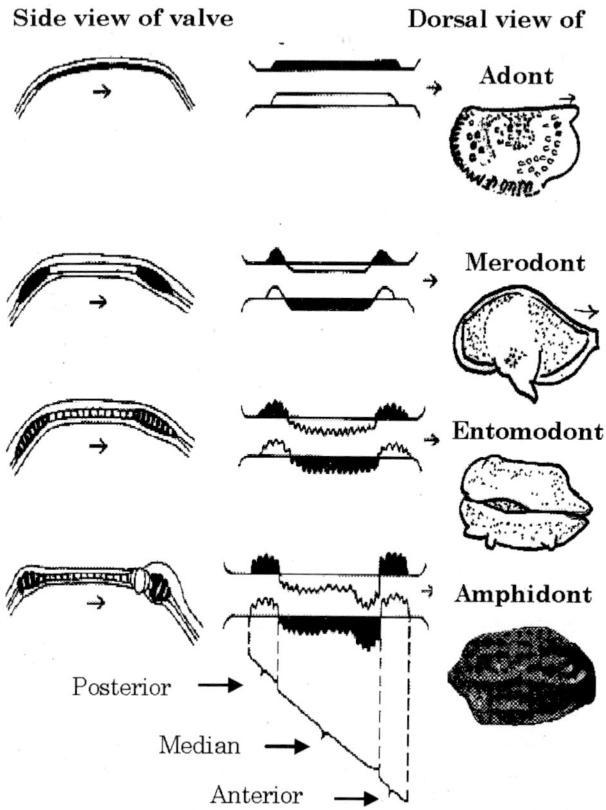

Description of the hinge

The simplest of all, lacks 'teeth and Socket', in general but may consist of 'grooves and sockets' instead, in the hinge zone of larger and stronger valves, respectively e.g., *Beyrichia*.

It has 3 basic hinge elements: anterior, median & posterior. The anterior & posterior elements show 'tooth and socket', while the median show 'grove and ridge' arrangement, e.g. *Cytheropteron*.

It differs by having a coarsely crenulated median element, e.g. *Cyprides*.

The first-part of the median element developed into a 'teeth and socket arrangement, e.g. *Schizocythere*.

Figure 6. 10 Types and descriptions of hinges (modified after Brasier, 1980)

Brasier (1980) has modified the types of hinges suggested by Van Morkhoven (1962) and Selvester-Bradley (1956), into four types (Fig. 6.10).

Adductor muscles and their impressions

The valves of a carapace are opened by an elastic, chitinous layer of outer lamella known as 'ligament' and closed by a series of transverse muscle, called adductor muscles, attached to inside of valves just at the antero-central region. After death of animal the muscles are destroyed by bacterial action and leave behind a 'cluster of muscle-scars' (Fig. 6.4). A central muscle scars area may be marked on the antero-central region. The cluster also includes mandibular- and fulcral-scars.

The adductor scars may vary in number, shape, size and in their overall pattern. They are important in the taxonomy.

6.4 ORIENTATION OF CARAPACE

Speciation requires differentiation of the left valve from the right and the anterior side from the posterior. A few guidelines (Triebels, 1941) are:

(a) Central muscle scar is situated a little ahead of the mid point in a valve;

(b) Median sulcus has its convex side directed towards the posterior end;

(c) Eye spot or eye tubercle is found situated at the anterior;

(d) The aerodynamic shape of the carapace;

(e) Carapace is the highest in the anterior and widest in the posterior end;

(f) Rostrum is always at the anterior end;

(g) In most cases, the posterior end of valve is pointed;

(h) Anterior part of the hinge is more developed;

(i) Marginal zone is more differentiated at the anterior; and

(j) Ventral side is often somewhat flattened, especially in benthic forms.

6.5 REPRODUCTION AND ONTOGENY

The reproduction in ostracoda takes place by the fusion of spermatozoa of unique size (up to ten times the length of males, produced by the testes) with the eggs produced by females. The eggs may be brooded inside the brood-pouches positioned posteriorly in the heteromorphs or released in the water between weeds or rock-fissures.

Ostracoda being are completely encased in a hard-covering, shed each time the animal adds to its body size. Therefore, a young ostracoda passes through eight larval stages known as 'instars' (Fig. 6.1). After the eighth molting stage the animal is referred to as adult. When the infant grows little beyond the space available, the existing carapace is shed and a new and larger one is secreted. Each molting of the carapace introduces an animal differing in form and size from that of the previous stage. Apart from the replacement of carapace in the second instar, the appearance and the degree of development of several anatomical and morphological feature from instar one to eighth may be enumerated as follows:

(i) Appearance of maxillae and furcae and full development of the later in the 3rd instar;

(ii) Appearance of the first thoracic leg in the 4th and full development in the 5th instar; the second in the in the 5th instar and full development in the 6th; and the third in the 6th instar and full development in the 7th;

(iii) Full development of mandibular and adductor muscles in the 7th instar; and

(iv) Full development of sexual organs; external genital structures e.g., brood-pouches; duplicature; hinge etc., takes place in the 8th instar.

6.6 SEXUAL DIMORPHISM

They have a well developed reproductive system. Sexual dimorphism is distinct as their male and female have carapaces of different shape and sizes. Their identification is important in taxonomy of the palaeocopids.

The female carapaces are called 'heteromorphs', which differ from the male and infant carapaces collectively called tecnomorphs. The females are inflated in the posterior region due to of presence of bulges known as, brood- or egg pouches. The sexual dimorphism is especially marked in fossil paleocopids having a special taxonomic value. In podocopans, it is less pronounced, but the males are generally longer and narrower; and the females have greater posterior inflations. Impressions of the reproductive parts are often found on the posterior inner surfaces of the valves.

Some freshwater environments may have the whole populations comprising of females only. Reproduction in such cases takes place without the participation of males. The phenomenon is known as 'parthenogenesis'. The females may outnumber male carapaces in certain marine environmental conditions.

6.7 CLASSIFICATION

Their classification may be attempted with two perspectives: (a) based on their soft parts; and (b) based on their morphological features, which may be observed in the fossils. The former approach suits the requirements of zoologists who classify them on the basis of 'structures of appendages', while the latter is useful to palaeontologists where that is to be based on phylogenetic developmental aspects that could be used in biostratigraphic zonation, paleoecology, paleoenvironment etc, especially, for the extinct Paleozoic texa that can be recognised only on the basis of carapace features. Scott and Smith (1951) and Van Morkhoven (1962) have discussed criteria utilized for taxonomic grouping within the Palaeozoic and post-Palaeozoic species. The correct orientation of carapace, and distinction of the anterior, posterior, dorsal and ventral sides are of profound importance in their. Some of them listed by Brasier (1980) in the order of approximate importance are: (i) basic carapace shape; (ii) muscle scar position and arrangement; (iii) degree of development and fusion of duplicature with the outer lamella; (iv) structure, shape size and arrangement of normal and marginal pore canals; (v) nature, location and degree of valve overlap; (vi) hinge elements; (vii) nature of sexual dimorphism, if present; (viii) nature of surface sculpture, and presence of eye spots; and (ix) nature of marginal zone; and (x) forms of selvage and flanges.

Class Ostracoda: Small crustaceans; unsegmented body, weakly differentiated into cephalon and thorax, completely enveloped by two valves, each of which having a completely crystallized outer lamella and partly crystallized inner lamella; the two valves are hinged dorsally, collectively known as carapace, without growth line, pierced by pores; carapace are shed and reformed periodically during the ontogeny to allow growth; generally have seven pairs of jointed appendages, four attached to the cephalon, and three to the thorax; sexes separated, but in certain cases male are not known; Cambrian to Recent.

The classification proposed by Van Morkhoven (1962), and later modified by Brasier (1980) is presented in the Table 6.1

6.8 ECOLOGY

As mentioned earlier the ostracods live in a very wide spectrum of aquatic environments viz. moist forest soils; small pond, lakes, streams, swamps and cave waters (even in hot-water springs); brackish; normal marine; and hypersaline water bodies.

In marine water, they are distributed horizontally in hyposaline to ultrahaline environments, (estuaries, shelf-seas, and lagoons), and vertically up to a depth of more than 3,000 m. However, the shallow, neritic, benthic habitat is the commonest.

They may be benthic or pelagic throughout their life cycle. The benthic texa occupy fresh as well as marine habitats. The freshwater forms have smooth, thin, weakly calcified carapaces of small and simple bean shape; while the marine are larger, thick, highly calcified with variable complex morphology. The freshwater forms spend most of the time of the life cycle in swimming several centimetres above the substrate.

The size, shape and sculpture of the benthic ostracoda broadly reflect the stability, grain size and pore size of the substrate on, or in which, they live. Crawling species dwell on soils, relatively fine-grained substrates tend to have a flattened ventral surface with weight distributing projections like alae or frills, keels and lateral spines. Those dwelling on coarser substrate in more turbulent, near shore habitats are commonly thick with coarse sculpture of ribs reticulations or robust spines housing sensory setae. The epifaunal (living within the top 0.5 mm of sediments) or burrowing species are small, smooth, elongated but with robust carapaces like *Kirthe*. The nektonic (pelagic) species spend most of their life-cycle in swimming in the oceans have their carapace more aerodynamic, smooth, thin-shelled and ovate (ventral margin is usually drawn outwards) to subcircular in lateral profile. They may have a projecting rostrum at the anterior end.

Table 6. 1 Previous classification of Ostracoda

Order, Suborders, Superfamilies, Families

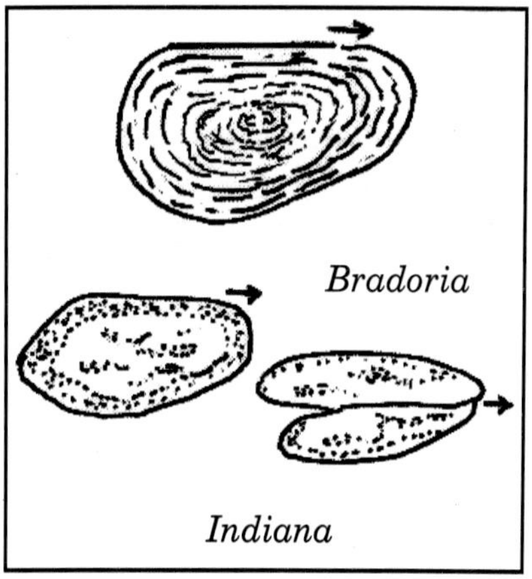

Order **Archaeocopida** (Cambrian to Ordovician)

Carapace thin; flexible, porous and of chitinous, phosphatic or weakly calcified material; hinge line straight, lacking in hinge elements; ventral margin convex, highest at posterior end, lacking in duplicature; no muscle scar, subcentral tubercles may indicate their original position; no sexual dimorphism; Shallow photic zone, as eye spots are present; found in Cambrian pelagic limestones in association with trilobites and paraconodonts; Cam.-Ord..

The Archaeocopida are may have been the ancestral stock of Leperditicopida.

Bradoria, a smooth or wrinkled, subquadrate form with pronounced eyespots; L. Cam.-M. Cam.
Indiana, a smooth form, elongate form lacking in eye spots; L. Cam.

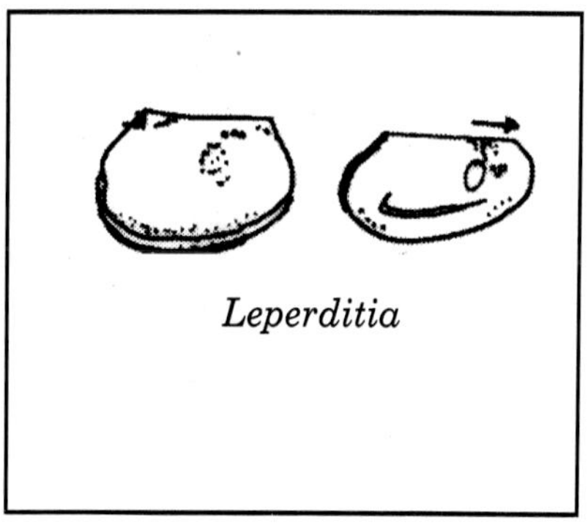

Order **Leperditicopida** (Late Cambrian to Devonian)

Thick calcified shell; valves are strongly unequal or subequal, the right valve overlapping the margin of the left one; large and distinguished muscle scars (5-30 mm long) comprising a cluster of more than 200 rounded scars; eye spot antero-dorsal; dorsal margin long and straight, may have cardinal angles; valve smooth; ventral margin convex undifferentiated and indicative of shallow-pelagic habitat; large size is suggestive of their carnivorous nature; developed form Archaeocopida forms in late Cambrian
They probably developed from archaeocopida in late Cambrian or early Ordovician, but diversification of species occurred in between Ordovician to Devonian.

Leperditia, a purse-shaped genus, smooth or punctate exterior, prominent eye spot and adduc.scars; L. Sil.-U. Dev.

Order **Palaeocopida** (Lower Ordovician to Middle Permian)

Palaeozoic forms recognizable on the basis of long straight dorsal margin and hinge line, often ending with cardinal angles; eye spot antero-dorsal (if present); sulci and lobes more developed towards dorsal margin, ventral margin generally convex, with frills, flanges, brood pouches or spines, especially in heteromorphs, a ventral lobe parallel to ventral margin is common; sexual dimorphism distinct.

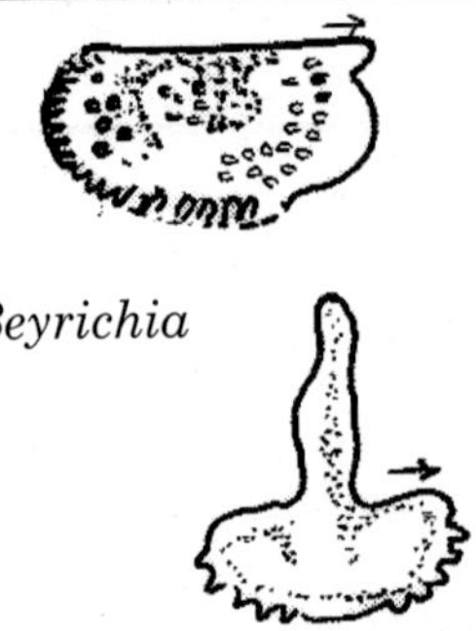

Beyrichia

Aechmina

Suborders **Beyrichiocopina** (L. Ordovician to M. Permian)

Right valve slightly larger, cardinal angle, lobes sulci and sexual dimorphism are pronounced; brood pouches or frills especially along the ventral border; hinge adont; swimmers near the bottom in shallow neritic zones; L. Ord.-M. Perm.

Beyrichia, a three lobed genus with pitted or granular exterior; heteromorphs has globular brood pouches; L. Silu.-M. Dev.

Aechmina, is characterised by a very large dorsal spine and the ventral margin by small spines; M. Ord.-L. Carb.

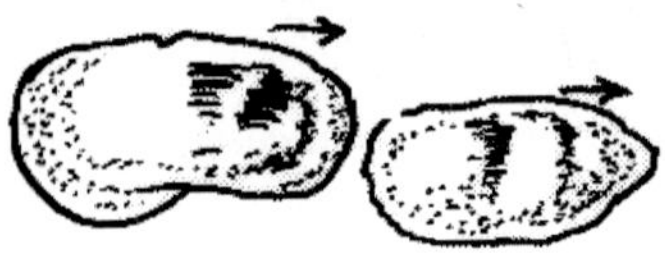

Kloedenella

Suborder **Kloedenellocopina** (L. Ord.-Perm.)

Valves subovate, right one slightly larger; rounded cardinal angles; less numerous lobes and sulci and sexual dimorphism less pronounced, as compared to Beyrichicopina; ventral margin less convex, may be straight or concave; crawling and burrowing habitat common.

Kloedenella, has larger left valve; heteromorphs has higher posterior than tecnomorph; Sil.-Dev.

Order **Podocopida** (Lower Ordovician to Recent)

Dorsal margin convex, less than the length of carapace; all types of hinge lines possible; eye spot may be found; ventral margin weakly convex straight or concave; adductor muscles positioned anteriorly; pointed end is generally posterior in lateral view, in dorsal and ventral view the posterior end is broader; major spines, tubercles and alae positioned posteriorly; crawl and burrow on marine sediments.

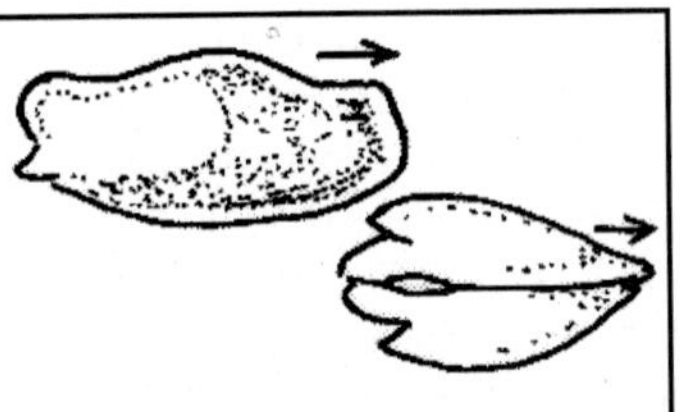

Healdia

Suborder **Metacopina** (? L. Ord. to M. Ord. to L. Cre.)

Left valve slightly larger; duplicature present, but narrow; exclusively marine, extinct.

Healdia, a rounded, smooth genus with backwardly directed shoulders, hinged by a crenulated bar and groove arrangement; Dev.-Per.

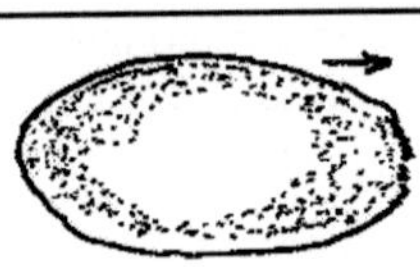

Cytherella

Suborder **Platycopina** (Jurassic to Recent)

Ovate, living ostracods, characterised by larger right valves, adont hinge and a cluster of 10-18 elongated adductor scars, weakly developed duplicature, and three pairs of thoracic appendages.

Cytherella, an ovate, smooth genus with more inflated posterior, as viewed in dorsally; Jura.-Rec.

Suborder **Podocopina** (Lower Ordovician to Recent)

Dorsal margin generally convex, hinge line adont, merodont, entomodont or amphidont; ventral margin concave, duplicature prominent; muscle scars fewer.

Podocopina were evolved through the Palaeocopida and Metacopina

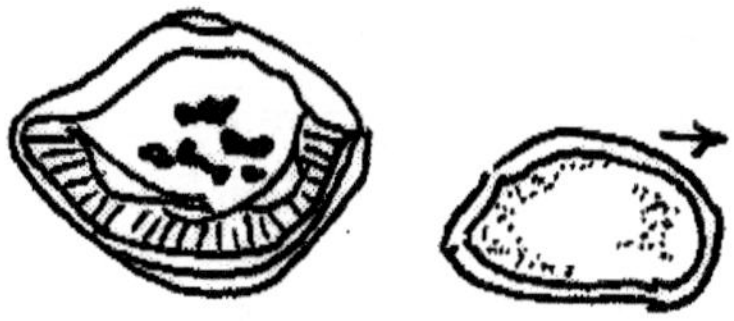

Bairdia

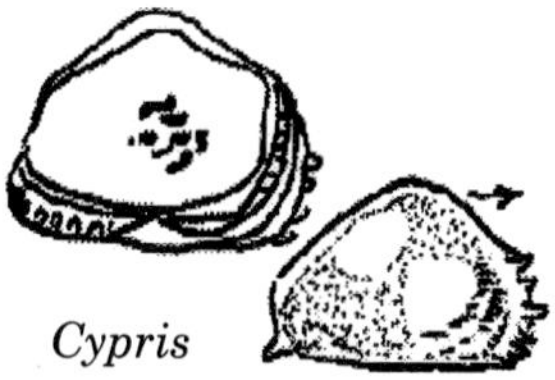

Cypris

Cypridea

Darwinula

Kirthe

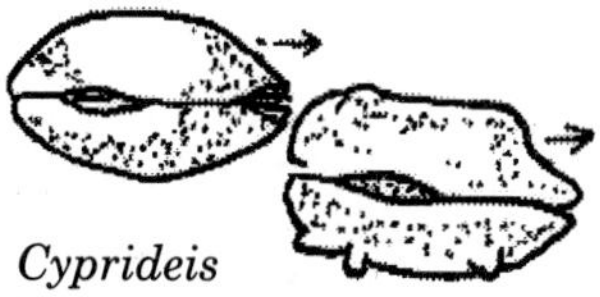

Cyprideis

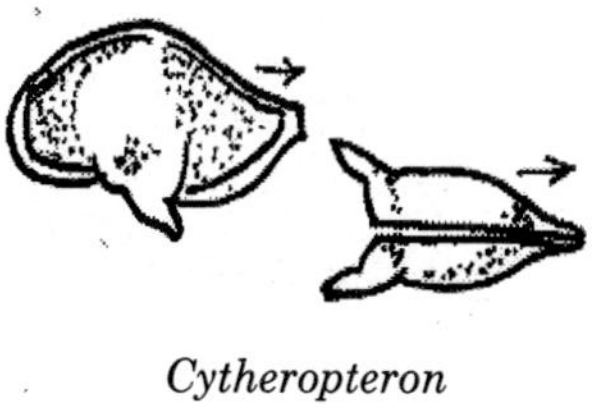

Cytheropteron

Superfamily **Bairdacea** (Lower Ordovician to Recent)

Carapace thick, smooth, dorsal margin strongly convex, anterior end blunt, posterior end pointed; duplicature wide; vestibulum prominent; 6-15 adductor elements; adont or merodont articulation

***Bairdia*,** the longest ranging ostracoda, left valve overlaps the right one; hinge adont; Ord.-Rec.

Superfamily **Cypridacea** (Devonian to Recent)

Freshwater taxa are common; have smooth thin, weakly calcified shells, poorly preserved; hinge adont or rarely merodont; six elongated, nearly aligned adductor scars, dorsal element larger then three anterior and two posterior ones; duplicature generally fused to the outer lamella; marginal zone narrow.

Cypris, the most widespread fresh-water, smooth, large (> 2.5 mm), subtriangular genus; ? Jura., Pleis.-Rec.

Caprice, large, extinct; surface pitted or pustulose; fresh-water to brackish water genus; dorsal and ventral margins relatively straight; antero-ventral margin beaked; hinge merodont; U. Jura.- L. Cret.

Superfamily **Darwinulacea** (? Ord., U. Car.-Rec)

A fresh-water genus with a very distinctive adductor muscle scar of four to five elements aligned in a vertical row; e.g., *Darwinula.*

Superfamily **Cytheracea** (M. Ord. -Rec.)

Variation in morphology, adductor muscle scars of 4-5 elements aligned in a vertical row; hinge generally merodont or amphidont; duplicature and marginal zones prominent; marginal pore canals branched.

Cyprideis, subovate; essentially smooth walled; entomodont hinge; brackish-water forms may bear hollow, tubercles; Mio.-Rec.

Limnocythere: fresh-water; carapace thin chitinous; hinge merodont; many marginal pore canals; hollow phenotypic tubercles in low salinities; Jura.-Rec.

Cytheropteron, a blind and soft-sediment dweller of variable marine depths is ovate/ subrhomboidal, merodont genus with distinctive alae and upturned caudal processes; L. Jura.-Rec. *Kirthe, Bathocyratina*

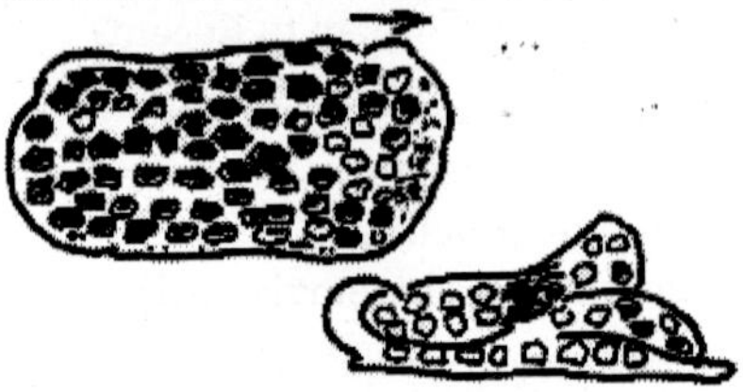

Bathociratina

Richreia

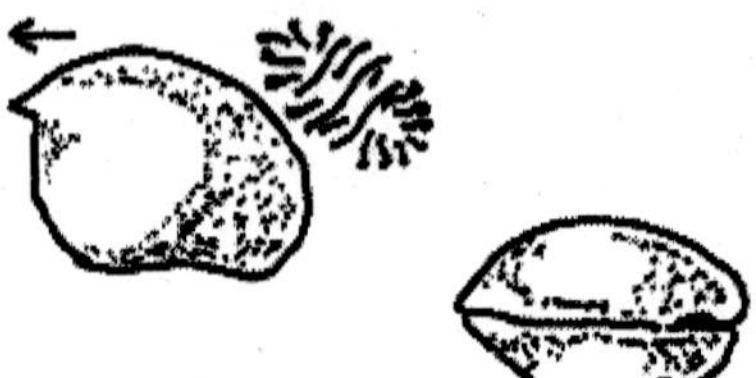

Entomoconchu

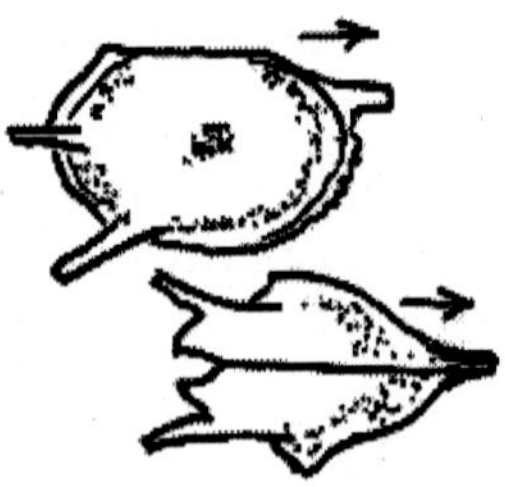

Thaumatocypris

Order **Myodocopida** (Ordovician to Recent)

Pelagic ostracods; provided with biramous antenna with a large basal segment for swimming; furcae large; dorsal margin convex or may be straight, adont; ventral margin convex may bear spines; duplicature narrow; muscle scar variable in number, of less taxonomic value; broader end posterior when viewed from dorsal side.

Suborder **Myodocopina** (Ordovician to Recent)

Several living and extinct superfamilies

Superfamily **Entomozoacea**

A Paleozoic group of oblong carapaces which bear concentric and longitudinal striae e.g., *Richteria*; (Silu.-Perm.)

Superfamily **Entomoconchacea** (Dev.-Carb.)

Large subangular ostracods with a siphonal gape at the posterior end and a distinctive pattern of muscle scar e.g., *Entomoconchus* (L. Carb.)

Superfamily **Thaumatocypridacea** (Mi. Jura.-Cret.)

Valves thin, weakly calcified; fossil forms are thicker with heavy ornamentation; both bear anterior spines e.g., *Thaumatocypris* (M. Jura. Rec.

Superfamily **Cypridinacea** (U. Cret.-Rec.)

Carapace ranging from Silurian to Recent have a prominent anterior rostrum; two stalked compound eyes and a median simple eye e.g., *Cypridina*.

Suborder **Cladocopina** (? Dev., Jura.-Rec.)

Differ from myodocopids in lacking eyes, heart and second and third thoracic appendages

Polycope, a well rounded genus of distinct cladocopine adductor muscle and without rostrum; a weak-swimmer that lives within substrate.

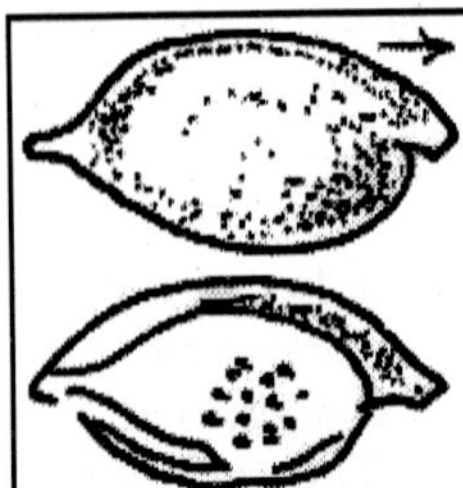

Cypridina

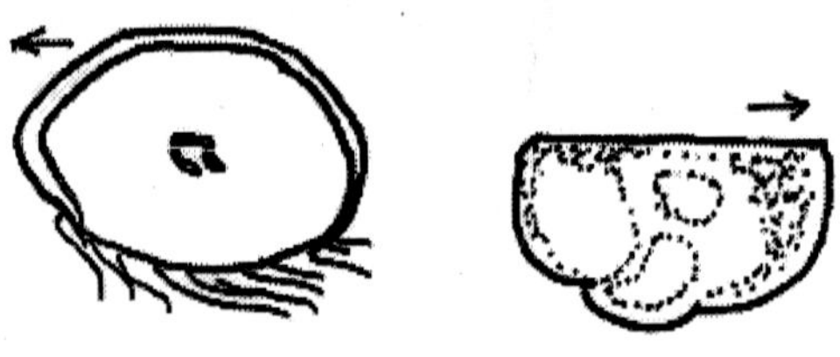

Polycope

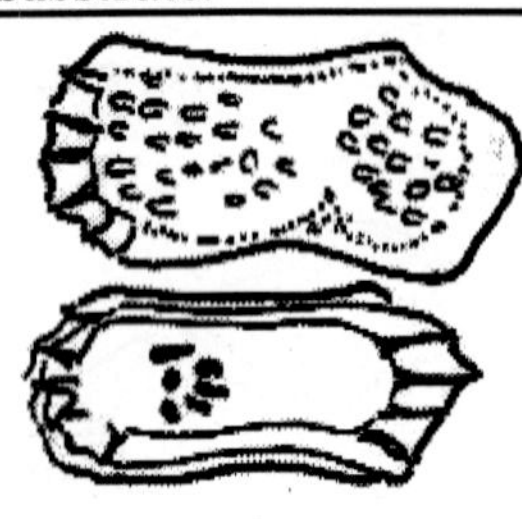

Beyrichia

Whatley *et al.*, 1993 (*vide* Armstrong and Brasier, 2005) divided Class Ostracoda into eight orders. The first four namely (Archaeocopida, Bradoriida, Leperditicopida and Eridostracoda) have been regarded as putative or presumed ostracoda as shown in the Table 6.2

Table 6.2 Putative Ostracoda Suborders (Armstrong and Brasier, 2005)

Status	Putative Ostracoda
 *Vestrogothia*	**Order Archaeocopida** (Cam. to Ord.) Carapace thin; flexible, porous, chitinous, phosphatic or weakly calcified material; hinge straight, no hinge elements; ventral margin convex, highest at posterior end, lacking in duplicature; no muscle scar, subcentral tubercles may indicate their original position; no sexual dimorphism; shallow photic zone, as eye spots are present; found in Cambrian pelagic limestones in association with trilobites and paraconodonts; Cam.- Ord. e.g., *Vestrogothia.*
 *Bradoria* 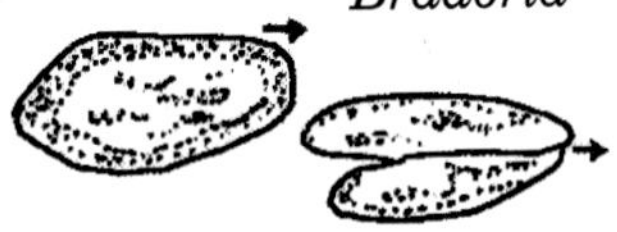 *Indiana*	**Order Bradoriida** (Cam. To L. Ord.) Small, adults are 1-18 mm long, smooth or wrinkled, subquadrate with prominent anterior-dorsal node. The appendages found in some fossils show that they do not belong to Crustaca. They include Bradoriida *sensu stricto* and the Phospatocopida *Bradoria*, a smooth or wrinkled, subquadrate form with pronounced eyespots; L. Cam.-M. Cam. *Indiana,* a smooth form, elongate form lacking in eye spots; L. Cam.
 *Lepeditia*	**Order Erodostracoda** Their carapaces have straight hinge-liner, curved ventral margin and concentric ridge sculpture. Their taxonomic position and rank of this order are controversial, as some scientist find that they are extinct group of marine branchiopod, while others consider them as Podocopida.
	Order Leperditicopida (L. Sil.-U. Dev.) Carapace large, well calcified, usually long (some up to 5 cm in length), straight hinge line and prominent eye tubercle, inner lamella non-calcified, complex muscle scar pattern with up to 200 subsidiary small scars. *Leperditina*, purse shaped, smooth or punctuate carapace with distinct tubercle and adductor-muscle scars.

Table 6.3 True Ostracoda Suborders (Armstrong and Brasier, 2005)

Sketches	Ostracoda Suborders
 *Beyrichia*	**Order Paleocopida** (L. Ord. to M. Perm.) Palaeozoic forms recognizable on the basis of long straight dorsal margin and hinge line, often ending with cardinal angles; eye spot antero-dorsal (if present); sulci and lobes more developed towards dorsal margin, ventral margin generally convex, with frills, flanges, brood pouches or spines, especially in hetero-morphs, a ventral lobe parallel to ventral margin is common; sexual dimorphism distinct. e. g., *Beyrichia.*
 *Propontocypris*	**Order Podocopida** (L. Ord. to Rec.) Dorsal margin convex, less than the length of carapace; all types of hinge lines possible; eye spot may be found ; ventral margin weakly convex straight or concave; adductor muscles positioned anteriorly; pointed end is generally posterior in lateral view, in dorsal and ventral view the posterior end is broader; major spines, tubercles and alae positioned posteriorly; crawl and burrow on marine sediments e.g., *Propontocypris.*
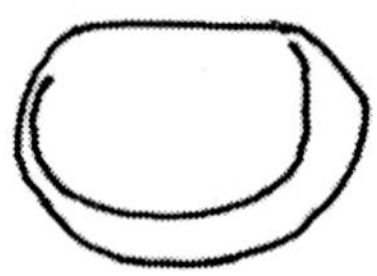 *Aparchites*	**Order Leiocopida** (Ord. to Perm.) They are superficially similar to paleocopids except lacking lobes and sulci. The carapace is inequivalved and adductor muscle-scar is rarely visible. *Aparchites* (L. to M. Ord.), oval, non-sulcate carapace, hinge line shorter than length cardinal angles are obtuse and the velar edge is smooth or tuberculate or spinose e.g., *Aparchites.*
 *Cypridina*	**Order Myodocopida** (Ord. to Rec.) Carapace weakly calcified with equal or unequal valves without valve overlap, the dorsal and ventral margins may be convex and the inner lamella is partially calcified. They are pelagic by habitat. *Cypridina* A typical myodocopid with a prominent anterior rostrum and rostral incisures e.g., *Cypridina.*
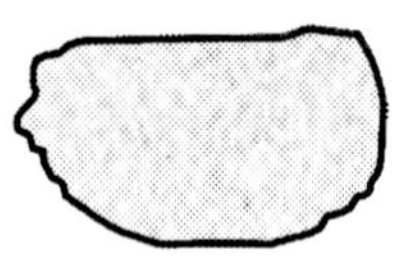 *Halocypris*	**Order Halocypridina** (Sil. To Rec.) Carapace almost entirely non-calcified, dorsal margin more or less straight and extends into a prominent rostrum. *Halocypris* a typical genus of the order, with a short rostrum.

Table 6.4 Classification based on carapace & adductor scars (Armstrong and Brasier, 2005)

	Subclass Podocopa
	Order Podocopida (Ord.-? Rec.)
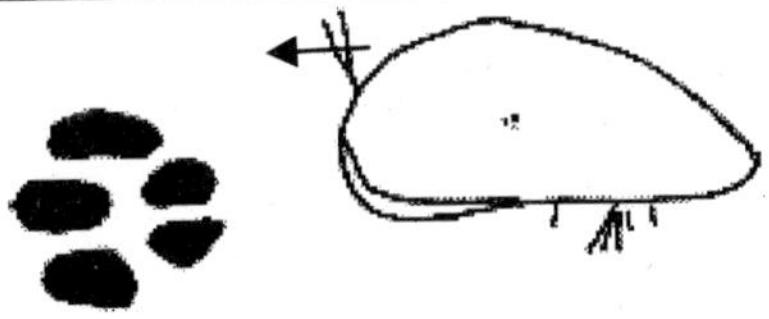	Suborder Cypridocopina Superfamilies: 3 Macrocypridoicea, Pontocypridoicea and Cypridoicea
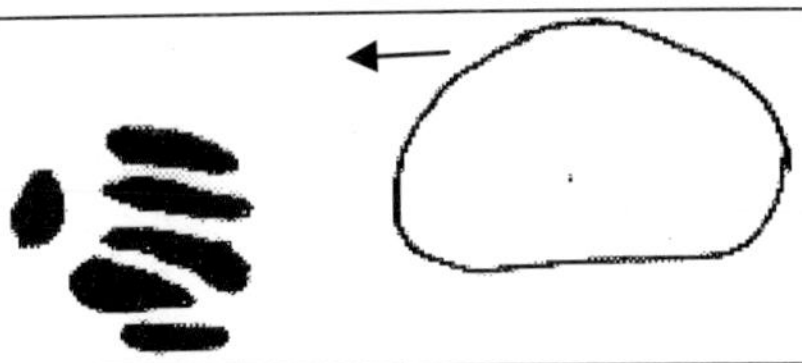	Suborder Cytherocopina Superfamilies: 2 Cytheroicea and Terrstricytheroicea
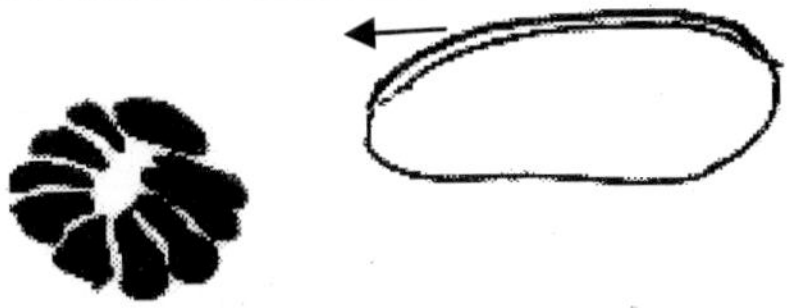	Suborder Darwinulocopina Superfamily: 1 Darwinuloicea
	Suborder Metacopina Superfamily: 1 Thlipsuracea
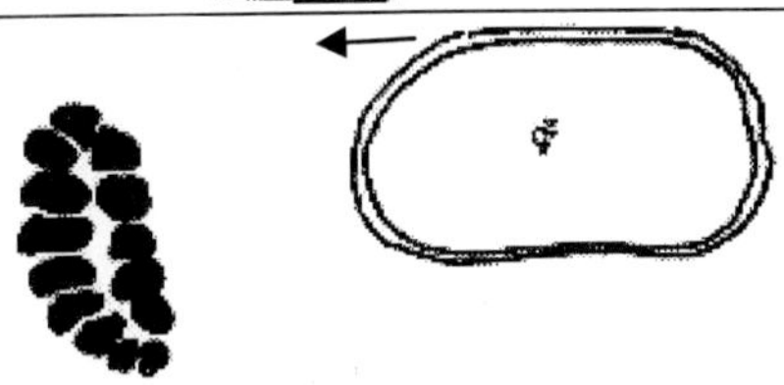	Suborder Platycopina Superfamilies: 2 Kloedenellacea and Cytherelloidea
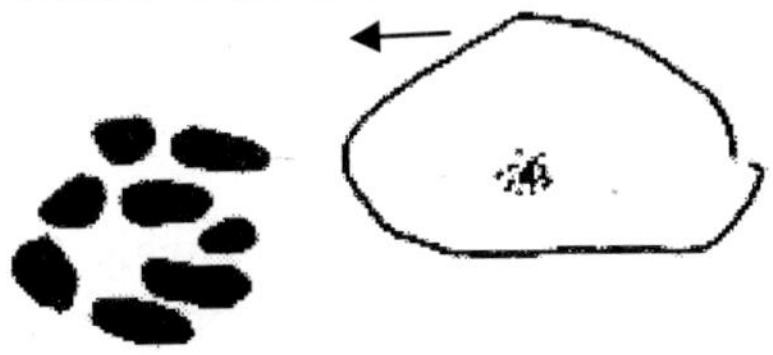	Suborder Podocopina Superfamilies: 4 Bairdiocypridacea, Bairdacea, Cypridacea and Cytheracea
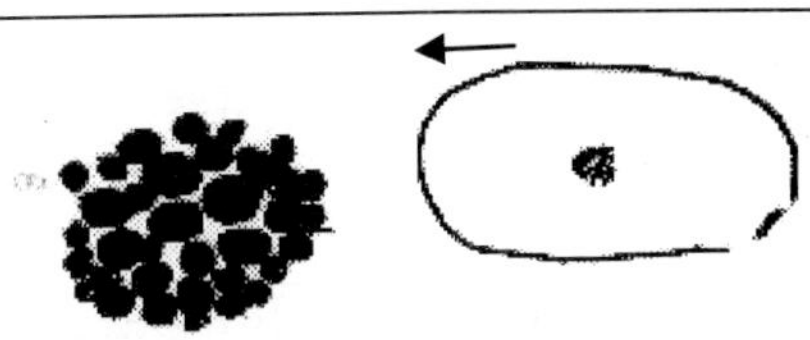	Suborder Sigillocopina Superfamily: 1 Siglloidea

Table 6.4 contd.

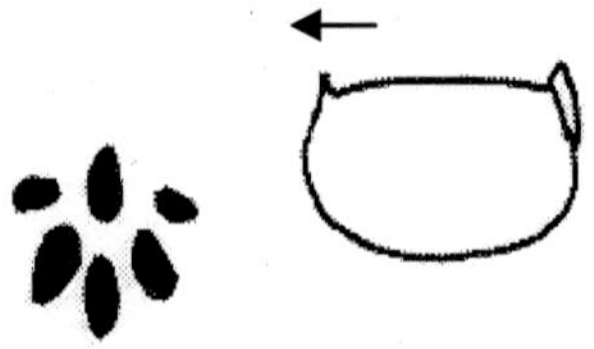	Order Paleocopida (Ord.-Trias. – Ter.) Superfamilies: 10 Barychilinacea, Beyrichiacea, Drepenellacea, Eurychilinacea, Holinacea, Kirkbyacea, Nodellacea, Primitiopsacea, Puncioicea and Tribolbinacea
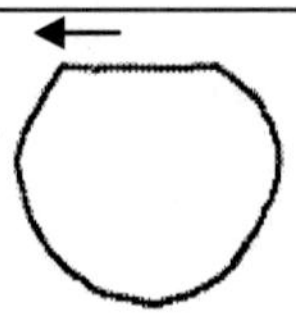	Order Lieocopa (Ord.-? Perm.) Superfamilies: 2 Aparchitacea and Parapachitacea
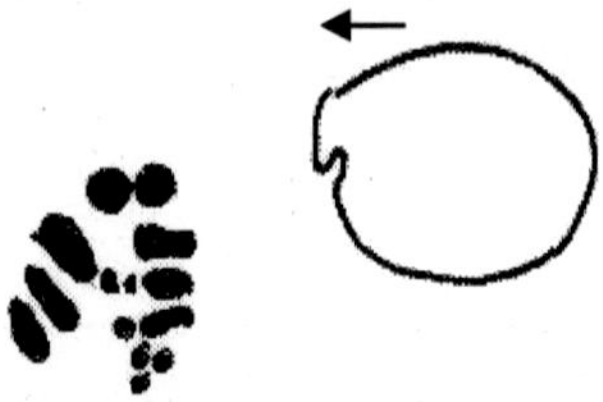	Subclass Myodocopa Order Myodocopida (Ord. Rec.) Suborder Myodocopina Superfamilies: 4 Cypridinoicea, Cyclindroleberidoicea, Entomozoacea and sarsielloicea
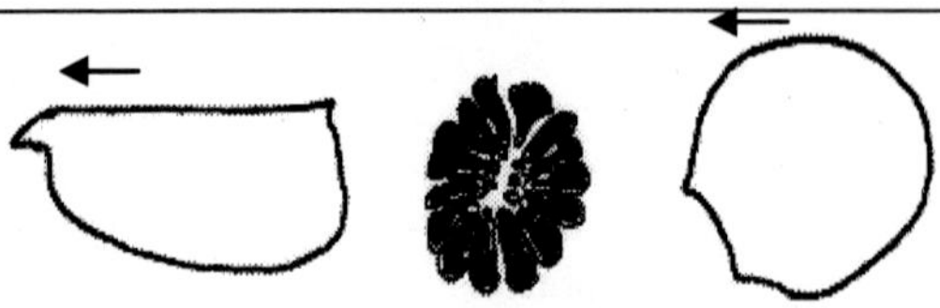	Order Halocyprida (Sil. Rec.) Superfamily: 2 Halocypridoicea and Thaumatocypridoicea
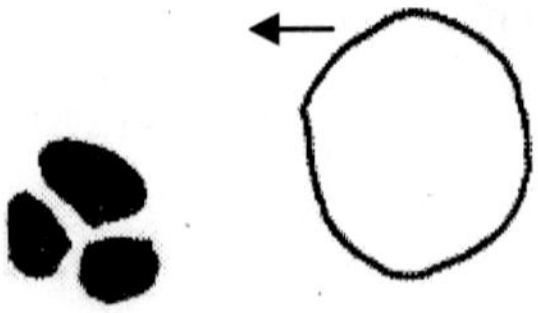	Suborder Cladocopina Superfamily: 1 Cladocopoicea

Most of the myodocopids and, several other thin-shelled species are rarely found as fossils. In paleoecology, ostracoda-genera provide less detailed information than their individual species (Von Morkhoven, 1962).

Their ecology assumed importance in late thirties and have been ranked next to foraminifera in deducing the paleoenvironment, especially of the oil bearing Tertiary rocks. The salient features are as follows:

(a) A large number of species, especially the post-Mesozoic ones still occur in the modern oceans;

(b) Various environments viz., moist forest-soils, freshwater ponds, lakes, paralic (hyposaline) water-bodies; normal salinity water-bodies; and hypersaline lagoons are characterized by their distinct assemblages;

(c) Variation in their morphology, in different geological periods is useful in dating, biostratigraphic correlation and interpretation of paleoenvironment;

(d) The ratio of open to closed carapaces helps in determining the rate of sedimentation;

(e) The degree of pyritization of fossilized valves points to the degree of reducing environment;

(g) The degree of deformation of valves is used to judge the degree of compaction of rock during palengenesis; and

(h) The phenotypic variations in carapace morphology helps in delineating the 'outer bay' and 'inner bay' facies in estuarine sedimentaries.

A large number of species can tolerate great fluctuations in environmental conditions. However, several other groups are restricted to well-defined biotopes and susceptible to slight changes in the environmental factors discussed below.

Salinity

It influences their ecology remarkably and is characteristic by the specific assemblages from freshwater to hypersaline conditions. Many of living species have very long geological record, therefore, their presence help in deducing the paleosalinities. The morphology of the carapace is influenced by salinity, as discussed below. Ramane (1934) found that: (a) the number of marine species decrease regularly down to a salinity of about 10‰ but shows an abrupt drop in the less-saline waters; (b) the number of freshwater species diminishes very quickly when the salinity surpasses 3‰; and (c) in salinities in between 3-10‰, ostracod species are not numerous.

Brasier (1980) recognised three main salinity assemblages: freshwater (>5‰), brackish water (0.5-30‰), and marine water (30-40‰). The hypersaline assemblage consists of marine and brackish water forms.

The salinity influences the ecology of Ostracoda is shown in Table 6.5.

Depth

Depth is a very important ecological factor. It may influence microorganisms because of depth related changes that occur in other environmental parameters, such as temperature, pressure etc. It is inversely proportional to temperature, light intensity, oxygen and directly proportional to pressure, and carbonate solubility. Due to the depth variation, the benthic assemblage of littoral, neritic, bathyal and abyssal environment vary considerably. Various depth related aspects of the environment are discussed in 'Chapter 3'.

The influence of depth on ostracoda in different environments is discussed here:

Freshwater They live in shallow pond, lakes and rivers. The depth zonation is noticed only in the deep part of large lakes. Two important forms of deep freshwater are *Candona* and *Cytherissa*.

Brackish water Brackish water environments, as a rule are shallow and very little can be said about the 'depth distribution' of species, confined to these habitats.

Marine water Ostracoda occur in practically all the marine environments ranging from very shallow tidal pools to depth exceeding 300 m but the largest number of them thrives in shallow neritic zones.

A progressive increase in depth causes lowering of temperature, as a result of which, the solubility of carbon dioxide is increased. As such the ratio of CO_2 to O_2 also increases with depth because algae can not photosynthesis below the photic zone, although the animals continue to respire. This leads to decrease in pH with depth from about 8.2 to as low as 7.0. There are distinct depth assemblages in both benthic and pelagic forms. In the latter is characterised by the following assemblages (Angel, 1969): (a) assemblage of rich diversity at a depth of less than 250 m; (b) an impoverished layer at 300-400 m; (c) rich assemblage at 450-625 m; and (d) rich assemblage at 750 m.

They show marked benthic zonation in 'inner-', 'outer-' and 'bathyal-abyssal' regions. The shelf-assemblage occur between 0 - 200 m depth and show brackish fauna. Dense populations occur in 'marginal marine environments', yet the highest diversity occurs in shallow waters. Those associated to marine algae occur within 100 m depth.

The coarse-substrate of shallow water is characterized by thick walled species, having eye spots, strong sculpture, amphidont hinge and branched marginal pore canals, e.g., *Quadracythere*. Deep water neritic-substrates of fine-grained sand-particles are inhabited by smooth, thin (often translucent) valves with weak hinges and no eyes e.g., *Kirthe*.

Bathyal and abyssal (1000-1500 m water depth) ostracod assemblage of constant temperature (3-4°C), salinity, pressure and the fine-substrate is neither abundant nor diverse but distinctive, as they are large (>1 mm), blind, highly sculptured and known as 'psychrospheric' fauna e.g., *Bythoceratina*.

Although they are cosmopolitan in distribution, they remain restricted to shallow water zones in polar latitudes.

Temperature

Ostracoda prefer a range of specific temperature for their living and reproduction. They inhabit waters of all temperatures; certain forms may have their optimum in cold water, others in warm water. Each species has a range of tolerance above and below the optimum and the range of this tolerance varies in different species, some up to 46°C in hot-water pools, e.g. *Heterocypris*.

There is a geographical or latitudinal temperature control of shallow water forms, which give rise to numerous localized (endemic) assemblages from the poles (temperature > 0° C) to the subtropics and tropics where the temperature may shoot up to 51° C. The tropical assemblages are more diverse than those of higher latitudes. It also controls the larval periods and maturation. The large body size of species of higher latitudes indicate their slower metabolism.

Table 6.5 Salinity influences on morphology of Ostracoda

Habitat (Salinity)	**Family/Genus**	**Carapace Morphology**	**Sketch**
Moist-soils	*Mesocypris*	Simple morphology, adont hinge	 *Mesocypris*
Freshwater (>0.5‰)			
Variable abundance & diversity, Cypridacea, Darwinulacea	Cyprididae *Limnocythere* *Cypris*	Smooth, thin shelled carapaces; adont (rarely merodont, never amphidont) hinge; no eye spots, branched marginal pore canals, caudal processes and hinge-ears	 *Cypris*
Brackish water (0.5-30‰)			
High abundance, low diversity, euryhaline, Cytheracea	Cytheridae *Cyprideis*	Thick shell, weak ornamentation, often punctated, merodont or modified amphidont hinge (rarely strongly heterodont) and remarkably low diversity	 *Cyprideis*
(5-10‰) Low diversity	Cytheridae *Cyprideis*	Development of hollow tubercles, an ecophenotypic indication	 *Cyprideis*
(5‰) Low abundance, high diversity	*Cyprideis*	Hollow tubercles more developed	 *Cyprideis*
Shelf Sea (30-40‰)			
Low abundance, high diversity, stenohaline	Cytheracea *Cytheropteron*	Carapace with rostrum, strong Cytheracea amphidont hinge, branching marginal pore canals, heavy, ornamentation well developed eye spot, sieve type normal pores, and/or complete fusion of inner and outer lamella	 *Cytheropteron*
Hypersaline Lagoons (>40‰)			
High abundance, low diversity	Cytheracea *Cyprideis*	Euryhaline forms, brackish water form (without tubercles), distinguished based on biota & sediments	 *Cyprideis*

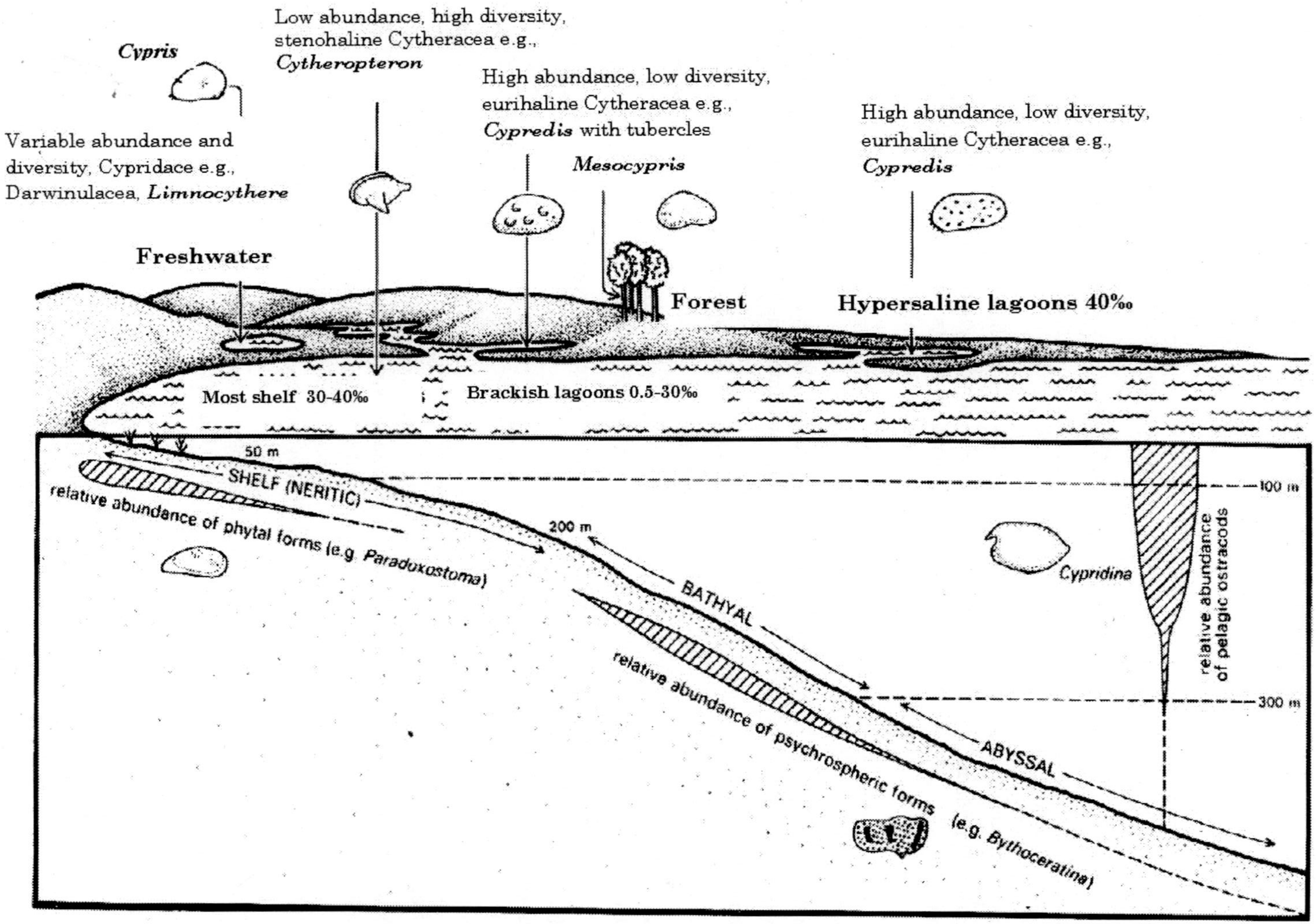

Figure 6. 11 Ecological distribution of Recent Ostracoda (after Brasier, 1980)

Substrate

The different substrates have their characteristic assemblages of benthic ostracoda:

(a) The ventral surface of species crawling on soft and relatively fine-grained mud is flattened. The carapace is provided with the weight distribution features like alae (e.g., *Cytheropteron*) frills, keels, or spines.

(b) The carapaces of the species living on coarse-grained substrate of near-shore or turbulent environments are coarse, robust, highly sculptured, and spinose e.g., *Quadracythere*.

(c) The freshwater species that burrow or live in pore spaces of sandy substrate are small, smooth but robust e.g., *Polycope*.

(d) The species that burrow silty and muddy substrates are long and smooth. Their carapaces are more streamlined e.g., *Kirthe*.

(d) The 'phytal ostracoda' of family Paradoxostomatidae suck juices of sea-grasses and algae, have smooth carapaces e.g., *Paradoxostoma*.

(e) Some of the pelagic species of current-upwelling zones are rich in phosphates and nitrates. The length of such smooth, thin shelled, and ovate-subcircular (in lateral profile) carapaces may go up to 30 mm e.g., *Gigantocypris*.

6.9 IMPORTANCE IN PETROLEUM EXPLORATION

The utility of ostracoda in high resolution biostratigraphy zonation, correlation, dating of rocks and interpretations of paleoenvironment, especially of Jurassic to Pleistocene rocks, is second only to the foraminifera. Several factors in the ecology of ostracoda may be used in study of the paleoenvironment of rocks of freshwater, brackish or marine origin.

The ecological factors may be used as tools in oil exploration mainly for identifying the source rocks for hydrocarbons in sedimentary basins. Oretli (1971) studied the data of some 25 oil fields and numerous well sections and, suggested some simple and rapid methods without knowledge of systematics are utilized. The basic approach is to determine the conditions of sedimentation based on shell characters and to interpret whether or not these were optimum for the generation of hydrocarbons. The organic matter may be converted in to hydrocarbon if the deposition of sediments takes place rapidly, in reducing aquatic conditions under the optimum pressure and temperature.

Some morphological characters reflect such ecological factors: (a) the carapace and valve ratio; (b) the degree of pyritization of the carapace; (c) the current index; (d) the compaction ratio based on the percentage of deformed carapaces; and (e) the preservation of carapaces.

Carapace and valve ratio

After the death of animal the adductor muscles of carapace disintegrate and the valves tend to spring open during the transportation of ostracoda. However, if they get buried earlier than opening of their valves the percentage of closed carapaces will be more than open valves. More is the number of closed carapaces higher will be the rate of sedimentation. Guha (1986) found that the Eocene sediments in Cambay basin with 100% closed carapaces were deposited rapidly while the Miocene sediments in Kutch Basin, having 80% closed-carapaces, in a relatively slower rate of sedimentation.

Degree of pyritization

The white or light brown colour of carapace of the optimum oxygen condition changes in to a shade of dark brown to black in the reducing conditions due to the accumulation of tiny pyrite crystals. This factor gives an access to determine the degree of reducing environment important for transformation of organic matter into hydrocarbons.

Current index

The current index is determined on the basis of percentage of adult and juvenile shells in a rich fauna. In an undisturbed fauna, both the adult and juvenile carapaces are expected. If the currents are active they may: (i) disturb the ratio of adult/juvenile carapaces; and (ii) create considerable aeration resulting into oxidation of organic matter. Usually a high current index is not favourable for generation of petroleum. Higher value of the index is met in the littoral, and low in the lagoonal assemblages.

Compaction ratio

Compaction of sediments is necessary for conversion of organic matter of sediments into hydrocarbons. This ratio may be determined directly on the basis of degree of deformation of carapaces or valves. It has been noticed that the thin carapaces easily get deformed in comparison to the thicker ones.

Preservation of carapaces

In the poorly preserved specimens the protective organic coating is not disintegrated. This indicates the incomplete conversion of organic matter into hydrocarbons.

Some other important aspects, helpful in oil exploration (Guha, 1986) are: (a) numerical relation between males and females; (b) relationship between right and left valves; (c) relationship between normal and dwarfed forms; (d) relation between smooth and ornamented forms; and (e) the diversity of species and their variation.

6.10 GEOLOGICAL DISTRIBUTION

It is now established that ostracoda of archaeocopid kind appeared together with the trilobites in the early Cambrian. But their origin, like other invertebrates is not yet known. Their swimming, filter-feeding practices appear to have been taken over in the late Cambrian by Leperditicopida. The literature on the living and extinct taxa reveals a new trends of evolution in ostracoda from the archaeocopids to the modern living forms:

(a) Increase in the curvature of the dorsal border from the straight-back ancestors;

(b) Modification of the hinge line from simple or adont to complex types with several teeth and sockets;

(c) Reduction in the number of instars;

(d) The calcification has become stringer;

(e) Elimination of second and modification of other appendages;

(f) Elimination of heart and degeneration of the circulatory system;

(g) Elimination of gills; and

(h) Development of complex sensory hairs on carapace.

In the early Ordovician, remarkable diversification occurred, attributable to the development of a variable nature of habitat due to a worldwide transgression. At this time appeared the first Palaeocopida (both Beyrichicopina and Kloedenellocopina), the first Podocopida (Bairdacea) and perhaps the first Myodocopida (Entomozoacea). The palaeocopids enjoyed the maximum population and diversity in the Ordovician period.

Their genetic diversity decreased from then until their complete apparent extinction in the Permian period. The Paleozoic ostracoda of great varieties occurred during Devonian and Carboniferous. The Podocopida were as many as they were in the Jurassic period accompanied by more myodocopids. During this time only the first freshwater ostracod (Cypridacea) appeared. In the late Devonian, time the leperditicopids and numerous other Paleozoic genera became extinct. A few new genera were added but they also declined and became extinct in the late Jurassic.

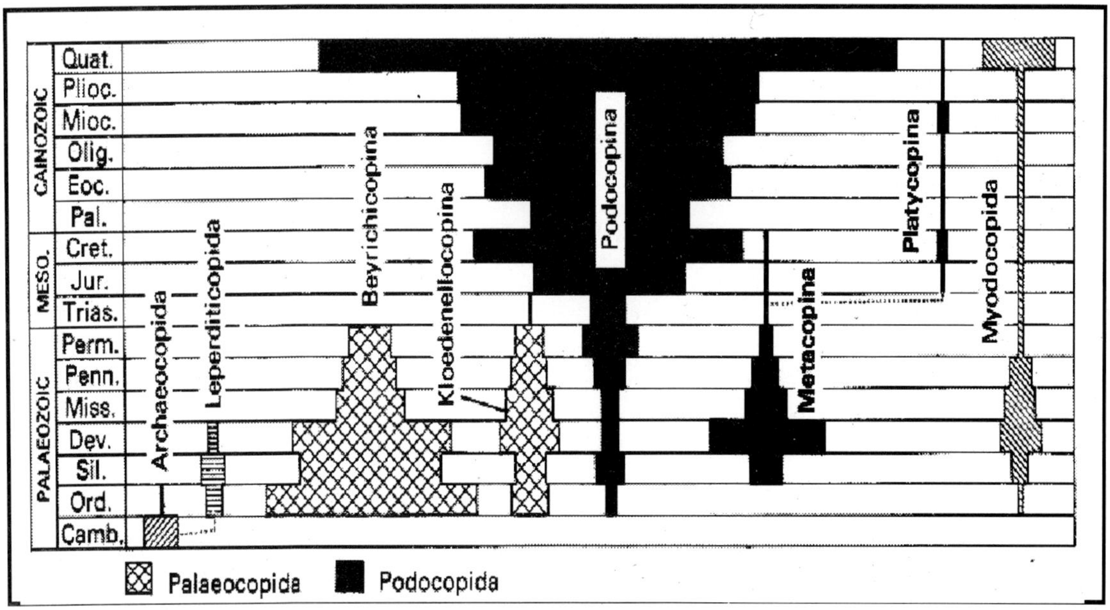

Figure 6.12 Diversity of Ostracoda taxa through time. Width of bars corresponds to the number of genera (modified after: Benson *et al*, 1961, in Braiser, 1982).

From the Triassic onwards, the value of ostracoda as paleoecological indicators became remarkable. In the start of Triassic, however, all the Palaeocopida suddenly disappeared. Only a few podocopids (*Bairdia* and *Darwinula)* could survive the Permo-Triassic boundary. During this period the podocopids started dominating all the benthic forms as a result of which they greatly outnumbered the Metacopina in diversity. The Platycopina evolved from dwindling Metacopina stock at about this time. Many of the Jurassic cytheraceans have proved to be good index-fossils owing to their short geological ranges e.g., *Camptocythere*.

The Cretaceous assemblages are diverse and dominated by the cytheraceans which evolved amphidont hinge elements in the late Cretaceous times. The Paleocene has witnessed a great diversification in their assemblages. The Pleistocene to Recent taxa added the poorly calcified stock by the appearance of new myodocopids and cypridaceans. The psychrospheric texa occupying the present deep oceanic water zones are thought to have been evolved in the late Cretaceous times when the Atlantic Ocean was developed due to the sea floor spreading.

GLOSSARY OF SELECTED MORPHOLOGICAL TERMS

Adductor Scar. Fig. 6.4.

Alae or lateral prolongation. The weight distribution projections developed to gain stability in the flat ventral-surfaced, benthic ostracodes e.g., *Cytheropteron.*

Antero central region, Antero dorsal region, Antero ventral region. Fig. 6.2

Antero-ventral beak or rostrum. A marked, abrupt indentation on each valve on the antero-ventral margin of some of the extinct cyprids (Family Cyprididae).

Brood pouches or egg pouches. Pronounced ventral lobes found on the posterior end of female or hetromorph form of ostracoda e.g., *Beyrichia.*

Caudal processes. A nozzle like projection, drawn out from the posterior end in some species. A hole which occurs at the tip of it, help in the better circulation of water.

Central anterior region, Central dorsal region, Central Postero region, Central region, Dorsal central region, Dorsal ridge. Fig. 6.2

Dorsal spine. A long tapering spine extended from the central dorsal margin in some species of suborder Beyrichicopina e.g., *Aechmina* Fig. 6.8b

Duplicature. The thicker and calcified part of the edge of thin chitinous **inner lamella** in adult Podocopida and Myodocopida. This is secreted to strengthen the interlocking between valves.

Eye tubercle or eye spot. A raised and extraordinarily large, upward pointing part of valve, situated below the antereodorsal hinge, found in some species e.g., *Indiana.*

Frills. Thickened, calcareous border on ventral margin, to provide more stability to the carapace (or a form of weight distribution feature) e.g., *Beyrichia.*

Hinge ears. Extension of the outer dorsal margin found in the larger valves of some species.

Inner lamella. The inner layer of the chitinous exoskeleton of a valve in ostracoda which is secreted by the epidermis cells of the nervous tissue. The margins of thin chitinous inner lamella are calcified into thicker **duplicature** in adult Podocopida and Myodocopida. This is secreted to strengthen the interlocking between valves.

Instars. One of several larval stages separated from other such stages by a **moult** or **molt**; generally eight in number.

Ligament. The two valves of an ostracod-carapace are joined and held together at the dorsal margin by a ligament.

Line of concrescence Fig. 6.4, **List.** Fig. 6.5, **Lobes.** e.g., ventral lobe Fig. 6.9, **Marginal zone.** Fig. 6.5, **Median sulcus.** Fig. 5.9.

Marginal pore canals. Numerous canal like structures in ostracoda which run across the marginal zone from the inner space to the outer world.

Moulting or molting. The process of in which some animals leave or shed their outer covers at each growth stage. Generally eight molting stages are known in ostracoda.

Muscle scars. The points of attachment of adductor muscles at the antero-central position in the inside of a valve in ostracoda.

Mandibles. The first pair of mouthpart appendages; usually short stout, unjointed structure forming bitting or chewing jaws.

Maxilla. An additional pair of mouth appendages located posterior to the mandibles.

Normal pores or lateral pores. Some additional pores, piercing the lateral surface of the outer lamella, in ostracoda.

Outer lamella. An outer and calcified layer of the chitinous exoskeleton of ostracoda (not found in some myodocopids).

Outer margin of carapace. Fig. 6.7, **Outline of carapace.** Fig. 6.7, **Outline of successive instars.** Fig. 6.1.

Parthenogenesis. A kind of asexual multiplication in which the ovum develops into a new individual without fertilization.

Postero-central region. Fig. 6.2, **Postero-dorsal region.** Fig. 6.2, **Postero-ventral region.** Fig. 6.2

Psychrospheric fauna. A distinctive ostracod assemblage occurring in bathyal and abyssal zones in nearly constant temperature, 3-4°C; salinity, pressure and the fine-substrate. It is neither abundant nor diverse but large (>1 mm), blind, and highly sculptured (p.11).

Ventro-dorsal region. Fig. 6.2

Posterior spines. Two or more calcareous spines found at the postero-ventral angulations of some species e.g., *Cyprideis* and *Paracypreides.*

Selvage. An inward projection of the **duplicature** which seals the two valves when closed

Ventral lobe. Fig. 6.9.

Vestibulum or valve cavity. The space or cavity left between the **inner margin** of the **duplicature** and the **outer lamella**, which accommodates the soft body-parts of the animal.

CHAPTER 7

Conodonta

Contents

7.1. INTRODUCTION

The Conodonta are phosphetic microfossils having tooth-like structures. Their affinity and function is now believed to be part of the feeding apparatus of an extinct early worm-like (hag-fish) vertebrate *Odontogriphus*. They are found in marine deposits, commonly in black-shales associated with graptolites, radiolarians, fish-remains, brachiopods, cephalopods, trilobites and palaeocopid ostracods.

They have been subject of greater controversy despite of a very careful study for about a century. Their taxonomic status dwindled from worms to gastropods or even fishes. Regardless of their phylogenetic relationship they constitute an important groups of microfossils as guide fossil and stratigraphic markers due to their small-size, relative abundance, widespread distribution and distinctive evolutionary transformation from late Cambrian to Permian, with widest distribution in Ordovician times.

7.2 TECHNIQUES OF STUDY

Being tough in nature they may be separated from rocks using weak organic acids (acetic, formic, or monochloric) supplemented by other physical methods.

Small in size (64 to 200 µm), they may be studied at a magnification of 150 or more. They can be sieved from finer materials and further concentrated by heavy liquid or ultrasonic techniques. The cleaned specimens can then be viewed using a reflected light microscope and mounted in slides at a magnification of 150 X or more. They can also be studied in thin-sections.

7.3 HISTORY OF RESEARCH ON ORIGIN AND AFFINITY

The name 'conodont' was coined by a Russian, C. H. Pander in 1856, who worked on Silurian fish fossils with an assumption that the ancient conodontophorid animal had a soft bilaterally symmetrical body with uncertain mode of life (benthic/nektonic or combined).

Despite being widespread and abundant their biological affinities were not known until the nineteen-eighties. Fay (1952) considered them a type of fish-like chordate, as they occur together with scales and teeth of Paleozoic fishes. Jones (1956) linked their affinity to worm-like invertebrate that presumably used the 'set of conodonts' for gasping rather than tearing or masticating food but others scientists gave different opinions the relation of conodont elements with gastropoda, lamellebrachia, annelida, chordate, arthropoda etc.

Ulrich and Bassler (1926) while describing several new species from North America recognised their biostratigraphic usefulness.

Schmidt (1934) discovered and Scott (1934) described a 'set of conodont elements' preserved together on the same bedding plain and opined that the individual elements were in life held in pairs (termed an apparatus) often likened to mouth parts (Fig. 8.1).

From the 1960's onwards conodonts have been used as one of the most important biostratigraphic tools available in Palaeozoic and Triassic rocks. Discovery of *Odontogriphus* a flat worm-like animal (Conway-Morris, 1976) showed that they belonged to an animal that had a flat, soft and gelatinous body with annular structures like earthworms as shown in Fig. 7.1.

Briggs *et al.* (1983) discovered a 4 cm long impression of worm-like animal with a terminal tail-fin that internally possessed conodont-elements in its proximal part (close to mouth) in Caroboniferous rocks.

More impressions of jaw-less animals (Aldrige *et al.*, 1983) indicated that the animal named *Clydagnathus* may be assigned to a new and separate Phylum Conodonta with animals having eyes, optic-capsule and possibly a notochord also. Some of their common genera are *Distacodus, Acodus, Prioniodus, Lonchodina* and *Lonchodus*. Their elements are found beneath the head region of the animal indicating their probable function as a 'complex feeding basket' helping in feeding, in capturing preys and cutting them into pieces.

Another group of phosphatic microfossils that bear resemblance with Conotonda is Scolecodonta, which is chitinous, jet black with dull lustre, without a basal bony-ridge and occur in shallow marine sediments from Ordovician to Recent.

7.4 BIOLOGY AND ASSEMBLAGES

Their concentric mode of growth shows that they were covered by soft-tissue and were secreted as internal structures, making them different from fish-denticles, gastropod radulae or the jaws of scolecodonts (annelids), which are held at the top of the secretary tissue.

Discovery of a fossils *Odontogriphus,* a flat worm-like animal, 60 mm long by Conway-Morris in 1976 showed that it belonged to an animal that had a flat, soft and gelatinous body with annular (like earthworm) form with U-shaped structure interpreted as lopophore. At the base of each tentacle of the animal is a compressed 'simple cone'. An assemblage found by Landing in 1977 showed that the conodonts were grasping mechanism rather than filtering organs.

Lelton and Scott (in Brasier, 1980) found numerous flattened cigar-shaped, 75 long and 15 mm wide fossil-structures having mouth, gut, notochord, fins showing chordate affinity 'conodont assemblage within gut that was considered as food-filtering system.

The worm like animal described by Briggs *et al.*, from the Carboniferous rocks showed that the conodont elements were the tooth-like feeding apparatus of a hagfish-like vertebrate. Accordingly, they represented some denticle of snake-fish like animals'. The co-occurrence of conodont elements in symmetrical pairs has allowed certain inferences to be made:

(i) The host animal probably had a bilateral symmetry;

(ii) The shape and arrangement of conodont elements in the apparatuses show that they were tooth-like feeding tools; and

(iii) Their SEM micrographs shows signs of wear on some elements revealing that the host animal probably produced only one-set of conodont elements in its life time.

Figure 7.1 Various conodont elements (apparatus) of *Odontogriphus*

Although they generally occur as discrete elements, many of their assemblages comprise upto 12-22 elements. Most of the well preserved assemblages recovered from Carboniferous rocks consist of one or more pairs of various genera arranged *in situ* on the bedding planes of carbonaceous shales. Such intact assemblages pose a problem in their taxonomy. One of the assemblages is shown in Fig. 7.1.

7.5 COMPOSITION

Their tooth-like structures are composed of calcium phosphate that is usually translucent and amber brown in colour.

X-ray crystallography shows that conodonts elements are composed of crystallites of apatite similar to the mineral francolite, embedded in organic matrix. Their phosphatic composition makes them durable/preservable even under adverse conditions of sedimentation/palingenesis. It also means that they were readily reworked/recycled into sediments of younger ages causing complications in biostratigraphic zonation.

7.6 MORPHOLOGY

Conodonts show either fibrous or lameller-structures (Fig. 7.2). They consist of basic single cusp, or large denticle, with or without auxillary-denticles, arranged on a sub-quadrate-to-subcircular base which usually has an escuteheon, or attachment cavity, on its side. Other forms consist of a series of cusp and denticles arranged on a bar-like base; which may be straight, curved, or re-curved sharply. Another large group consists of ridges, sulci, and nodes and a posterior serrate and denticulate blade, oriented normal to the platform.

7.7 CLASSIFICATION

Primarily the conodonts were grouped geometrically as:

(i) Coniform-cone-shaped consisting of elliptical base;

(ii) Ramiform-main cusp flanked on one or both the sides a row of teeth of similar/dissimilar size; and

(iii) Pectiniform-denticles on a platform bearing a number of radially directed processes.

By convention each of the conodont element was given a binomial name (genus and specific name) based on their shape. As a result a taxonomic problem arose when scientists attempted to classify elements of different names now found in a single apparatus.

Brasier (1980) recognised three groups of conodont elements on the basis of morphology:

a). Simple cone type: single tooth/horn like cusp pointed at the oral end and extended into a base at the aboral end. The anterior edge is convex while the posterior edge is concave, Fig. 7.2 (Cam.-Ord., Dev.). They were derived from paraconodonts and dominated in most of the Cambrian and Ordovician assemblages and gave rise to more complex elements of later times.

b). Blade type: ramiform conodonts formed by secretion of additional cusps, called denticles, borne on rounded bars or flattened blades on posterior or anterior side of the blade, with or without bent shaft that is commonly branched (Fig. 7.2) are collectively called 'compound element'. They developed from simple cones by the addition of denticles along anterior, posterior and lateral processes. A sub variety, bar type conodonts are conspicuous and common in many associations so it has only a few index-fossils whereas the blade type are good index fossils for a period from Silurian to Triassic.

c). Platform type: conodonts showing processes (especially at posterior end) that thickened into a basal ledge-like platform, sculptured on the oral side by ridges, tubercles and pits Fig. 7.2, They were probably evolved from the bar- and blade-types by development of the broad flanges into plates (excellent index fossils of Ordovician to Triassic).

Brasier (1980) also argued that "as the assemblage represent true biological unit it can have one name, which (according to the International Code of Binomial Nomenclature) must be that of the first described conodont element that it contains' and, therefore, classified the conodonts into:

Protoconodonts (U. Precambrian to L. Ordovician): Narrow, slightly curved simple, cones with a deep basal cavity and aborally-internally deposited lamellae with very thin wall and flattened area on the posterior edge, bound by two longitudinal carinae e.g., *Hertzina*.

Paraconodonts (M. Cambrian to Ordovician). Dark-brown to black elements having lamellae added aborally, with little overlap. Most of them are pixie-cap shaped with a deep basal cavity e.g., *Furnishina* that has a broad flat area on the anterior side of the cusp and a very flared base.

Euconodonts (L. Ordovician to U. Triassic). Basal body and a conodont proper with aboral-lateral and oral-lateral direction of lamellar growth. They may be divided in to simple, bar/blade and platform types of conodonts, which unfortunately not a biological distinction.

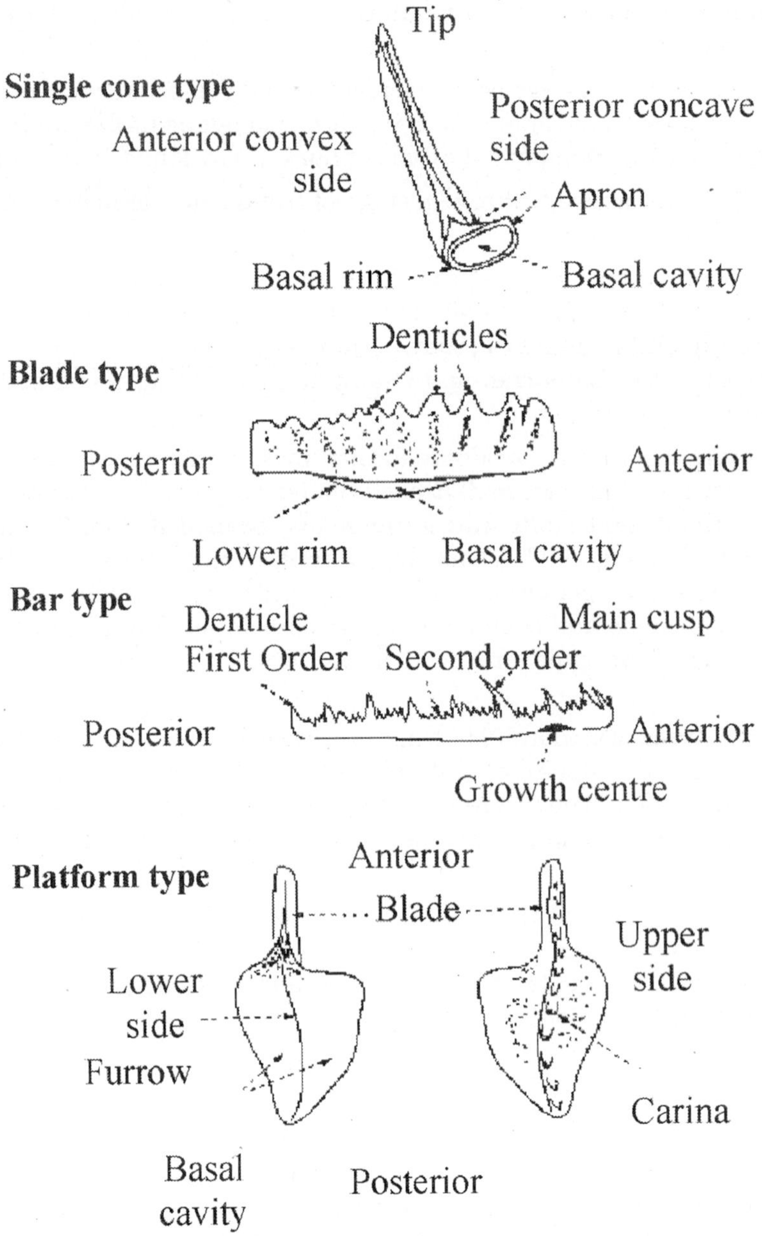

Figure 7.2 Morphological terminology (after Muller, 1978)

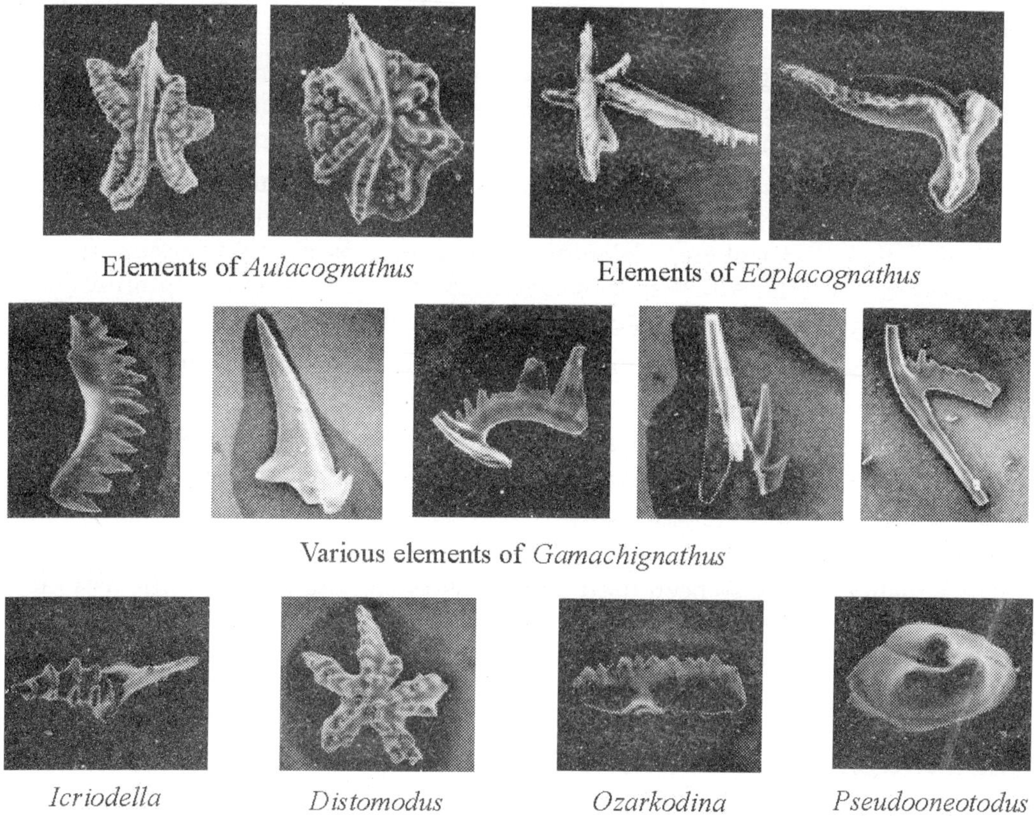

Figure 7.3. Various species of Conodonta and their elements

This category of 'compound element' conodont were further elaboration of the bar and blade model. Hass *et al.* (1962) recognised two distinct subgroups - Ploygnathidae and Idiognathidae. The former have basal cavity as tiny pit and the platform developed from swollen lateral edges of the posterior process, while the latter have large basal cavity, with lips extended into a platform that occupy much of the posterior end. Their main types are shown in Fig. 7.2.

Taxonomically, the conodonts were erected separately on the basis of their individual, distinctive, tooth-like object. Recently, they have been assigned to Phylum, Conodontophorida that has been divided into two Orders based on chemical and ultrastructure differences; eleven superfamilies recognised by reconstructing associations of individual elements into apparatuses; and into forty seven families based on morphological and element compositional differences. One hundred and eighty genera have been recognised.

The classification of conodonts is an un-natural one, as it is based on morphology only. Morphologically, four main groups of conodonts can be distinguished.

Some important conodonts are shown in Fig. 7.3.

7.8 ECOLOGY

The conodonts have great value in biostratigraphic classification and correlation of rocks from Ordovician to Permo-Triassic. Since all the conodont bearing animals are extinct their ecology has been worked out on the basis of fossils groups.

The animal that bore conodont-set lived in varied environments of deposition; the shale is by far the most dominating amongst them but they also occur in sandstone, carbonaceous shale, limestone, dolomite, chert etc. Since they occur in marine Ordovician-Permian rocks with fossils of contemporary animals like graptolites, it seems as though the animal preferred shallow, marine reducing conditions.

Study of their ecology has gained importance in view of its role in estimation of depth of deposition along with other environmental factors.

Change of colour of conodonts from pale amber to black shows the influence of temperature and time (in proportion to depth and duration of burial). Study of the chemical change in a nearby hydrocarbon resource may be made on the basis of the degree of change in colour.

Their lamellae have also been used in paleoastronomy to calculate the astronomical events such as number of days per year, tides or seasons etc. The data obtained have been correlated with the astronomical calculations based on rugose corals with many similarities.

7.9 GEOLOGICAL HISTORY AND DISTRIBUTION

By far they are most widespread and biostratigraphically important group of phosphatic microfossils as many of them are very good index fossils.

Their earliest species came from the Precambrian rocks of Siberia but they occur more commonly in Cambrian. Their diversity increased in the Ordovician and again during the Devonian. The conodont-bearing organism clearly survived the Permo-Triassic boundary extinctions but became extinct during the late Triassic.

Their extinction coincides with the diversification of dinoflagellates and first appearance of calcareous nannofossils. The most primitive conodonts were single coned that dominated early Ordovician. The first platform type conodonts also occur around this time. Their diversity and abundance, however, declined in the Silurian. During the early and mid Devonian their diversity gradually increased, reaching the maximum (acme) in the late Devonian. In the early Carboniferous, however, they became widespread and numerous but with low-diversity in late Carboniferous. In the Permian they were almost on the verge of extinction, but again flourished in early to middle Triassic only and disappeared in the late Triassic.

7.10 APPLICATIONS IN GEOLOGICAL STUDIES

The fact that conodonts are relatively common in rocks of Palaeozoic age, a period when other microfossil groups are either not present or scarce, has made them extremely useful stratigraphic tools. Together with acritarchs, chitinozoa and spores, conodonts are the primary microfossils available to palaeontologists working on Ordovician to Permo-Triassic strata.

As mentioned above the change in their colour from pale amber to black shows the influence of temperature and time (in proportion to depth and duration of burial). Studies on geochemical changes on their elements in a nearby hydrocarbon resource may be made on the basis of the degree of change in colour.

Studies on calculations on astronomical events such as number of days per year, tides or seasons etc., their lamellae are gaining importance. The data obtained have been correlated with the astronomical calculations based on rugose corals with many similarities.

Plate I **FORAMINIFERA**

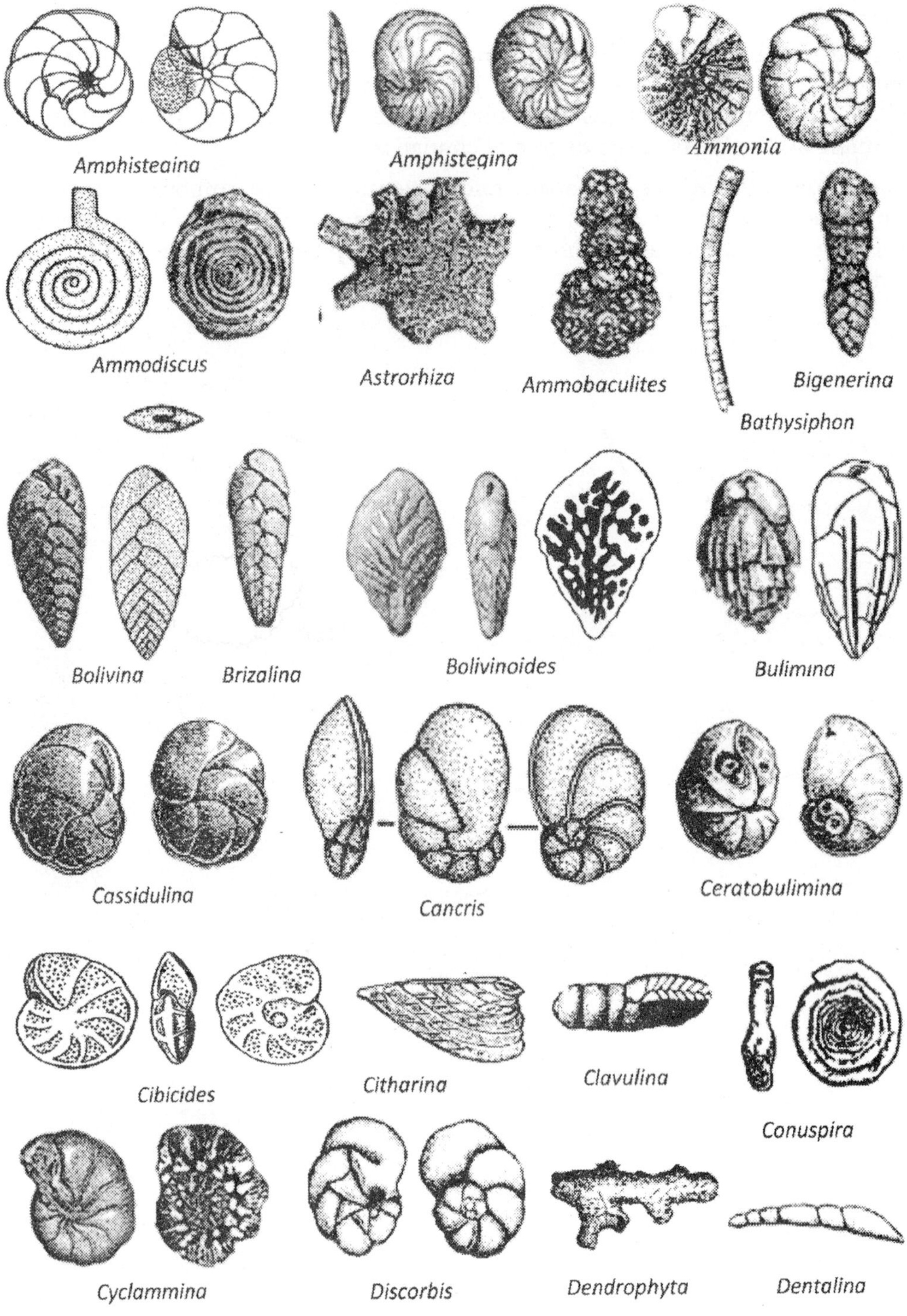

Plate II **FORAMINIFERA**

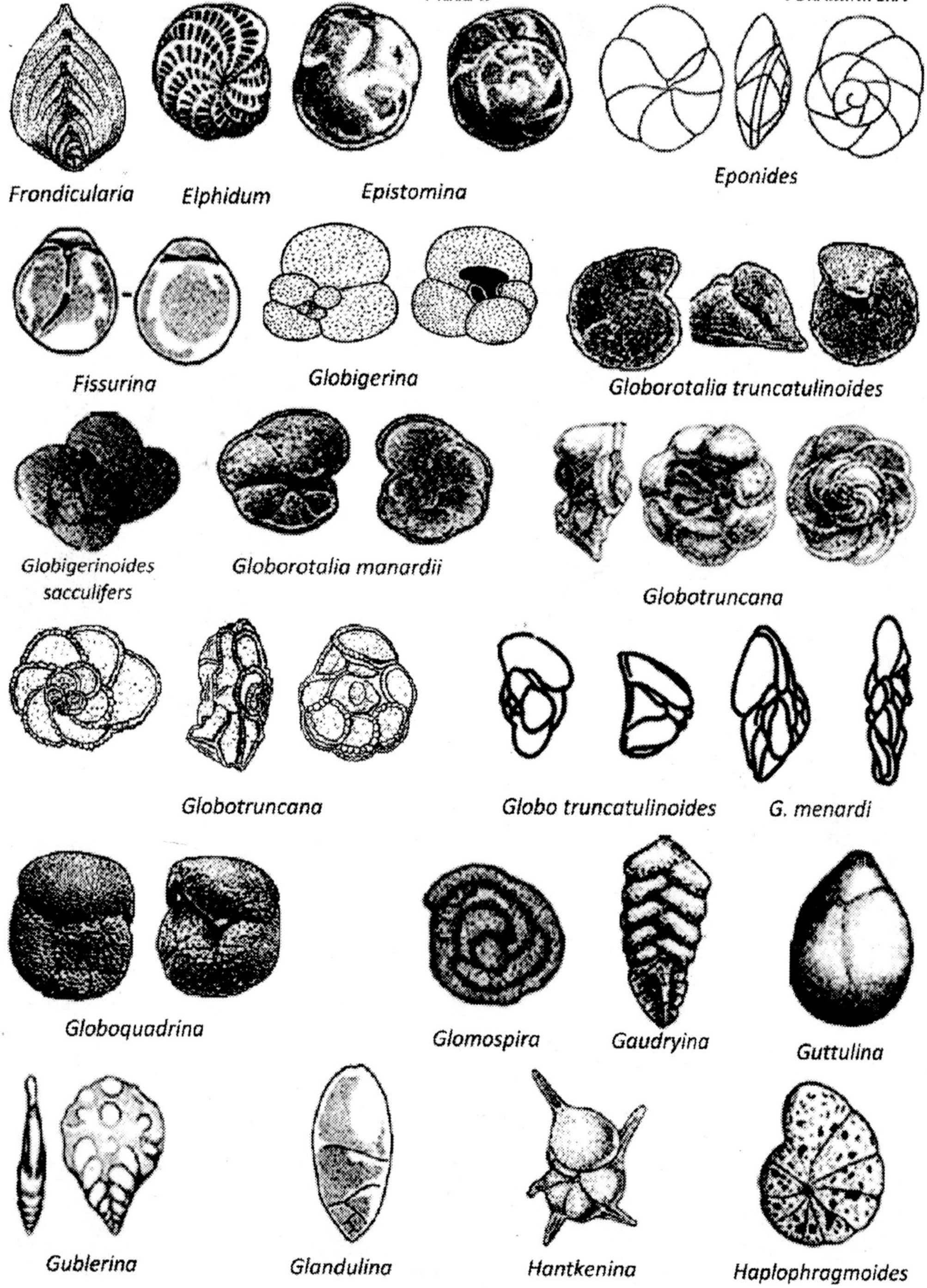

Plate III **FORAMINIFERA**

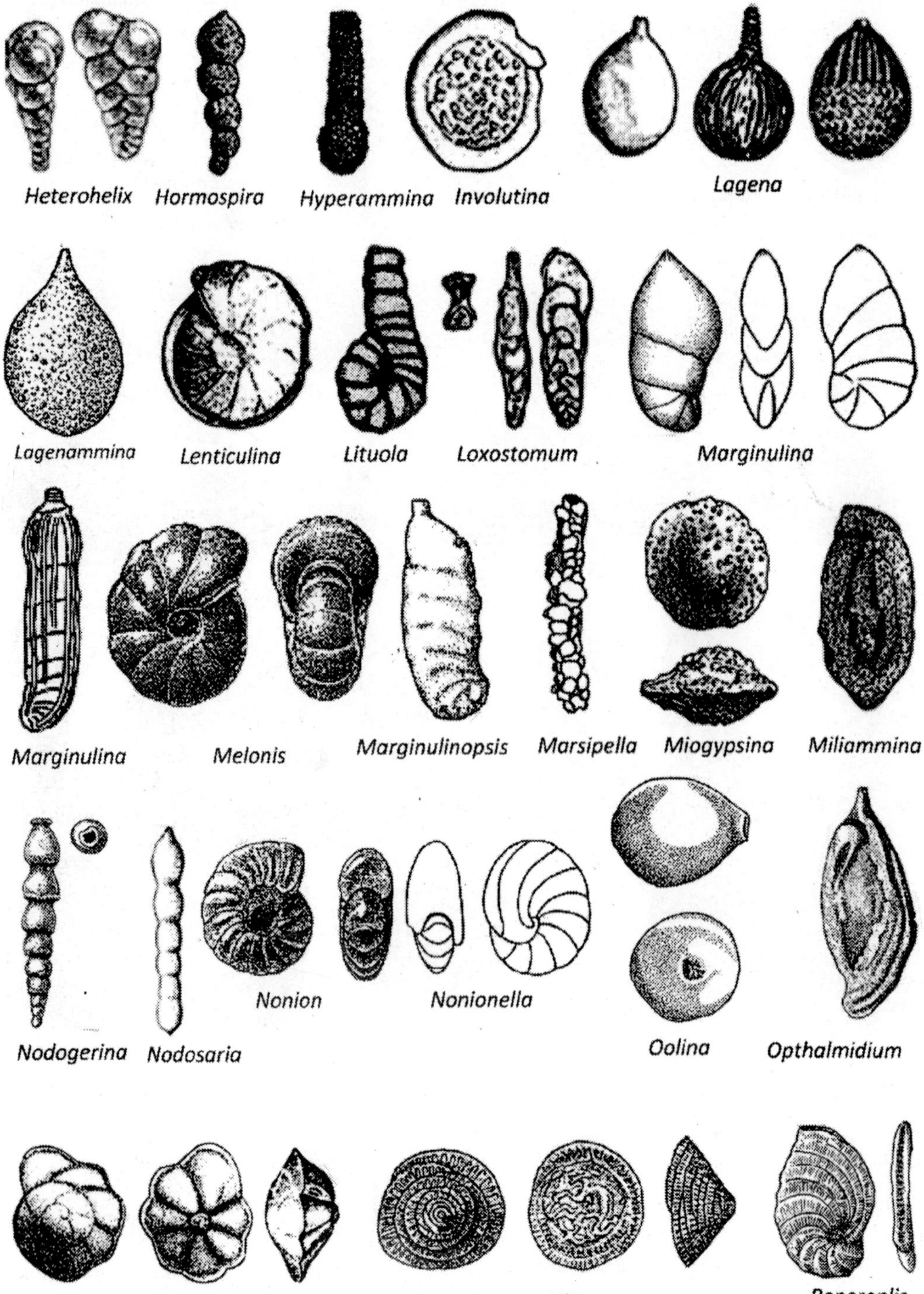

Plate IV **FORAMINIFERA**

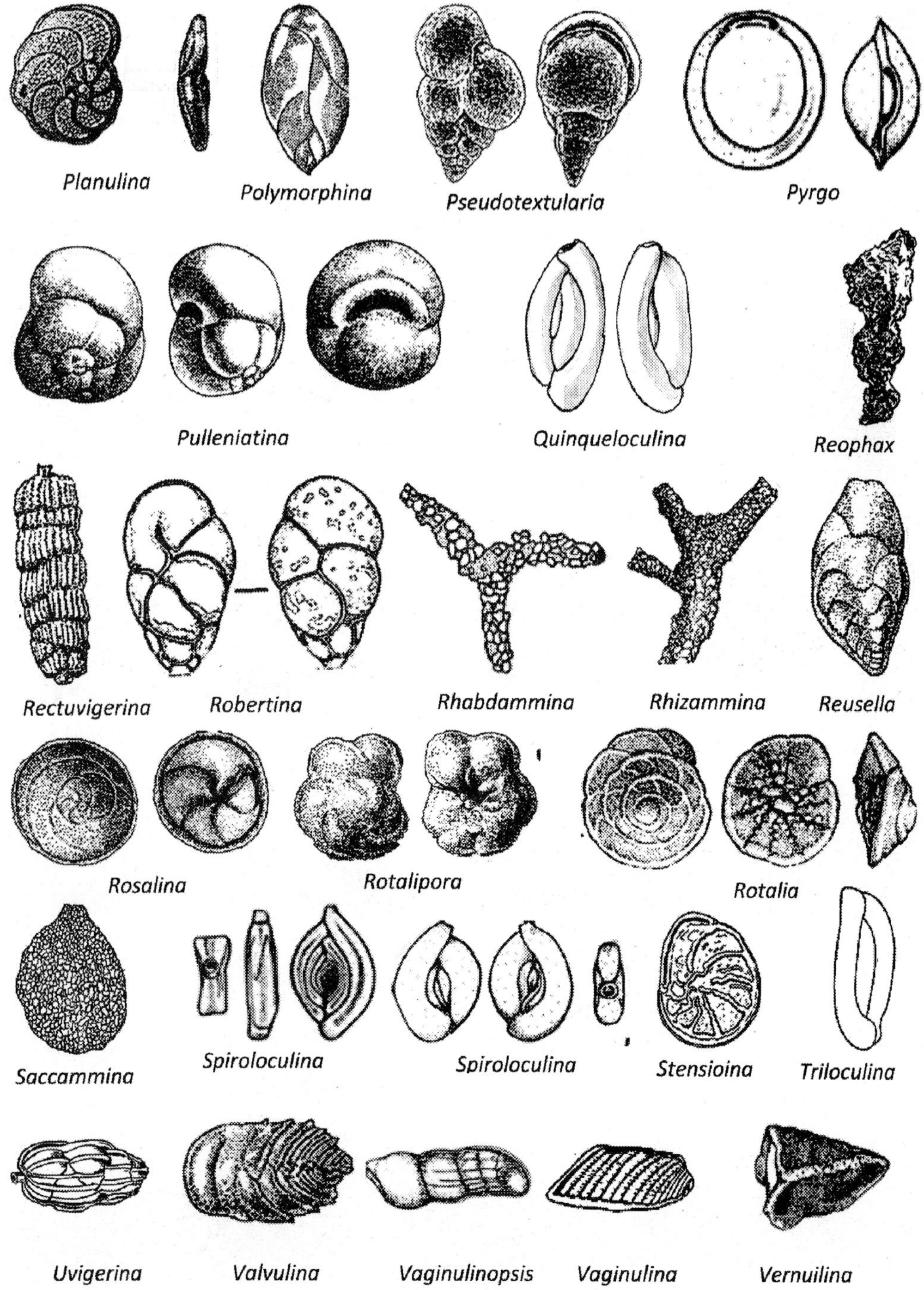

Plate V **DIATOMS**

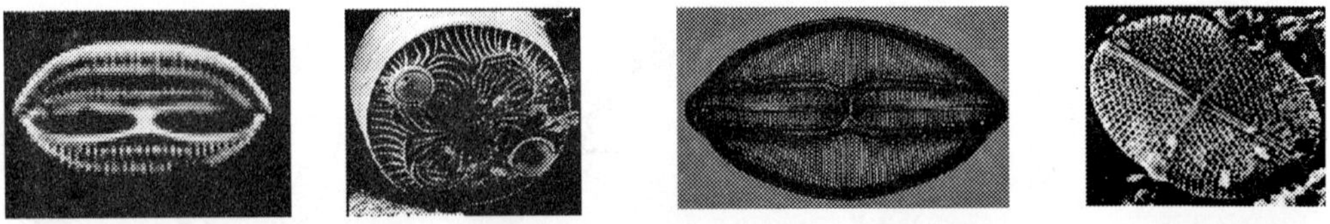

Diploneis *Auliscus* *Lyrella* *Cocconeis*

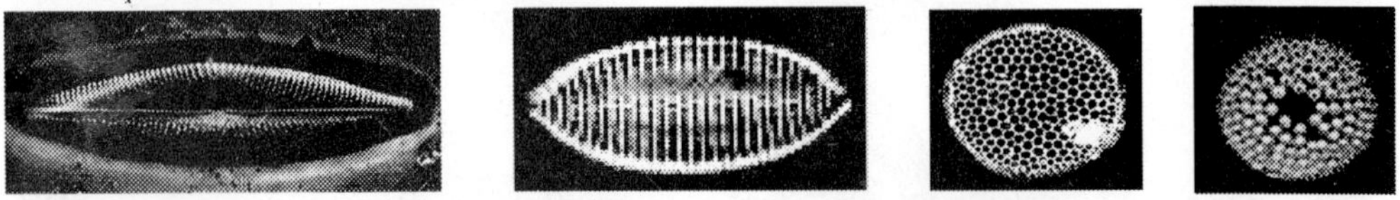

Dimeregramma sp. *Nitzschia punctata* *Coscinodiscus* *Actinocyclus*

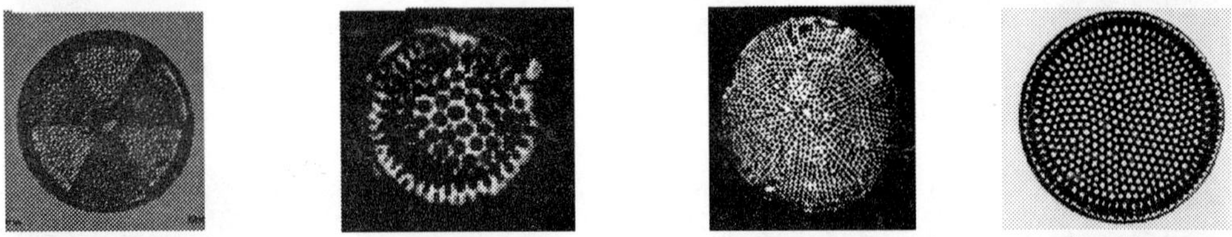

Actinoptychus *Thalassiosira* *Thalassiosira* *Azpeitia*

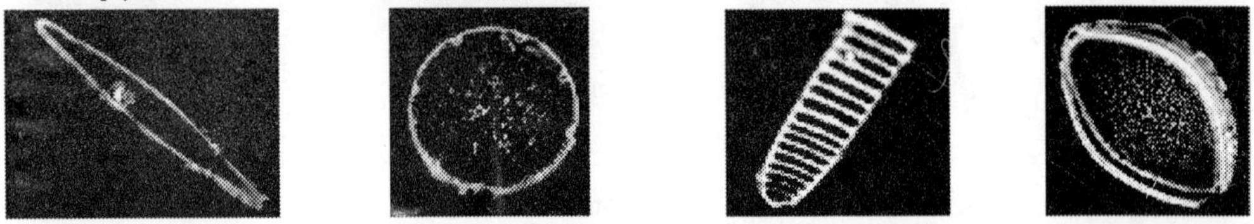

Nitzschia reinholdii *Fragilariopsis* *Hemidiscus*

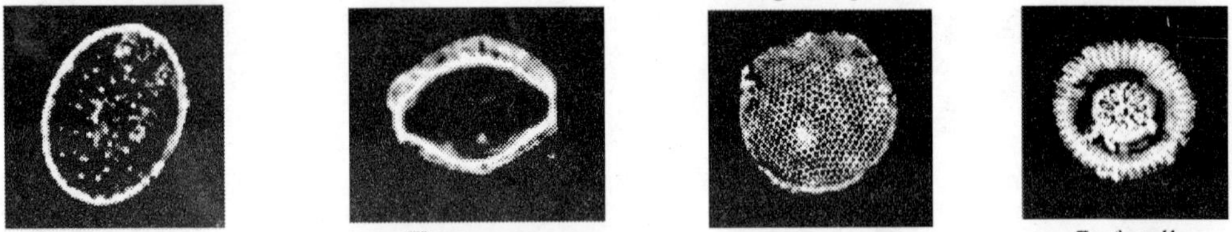

Chaetoceros *Chaetoceros* *Thalassiosira* *Cyclotella*

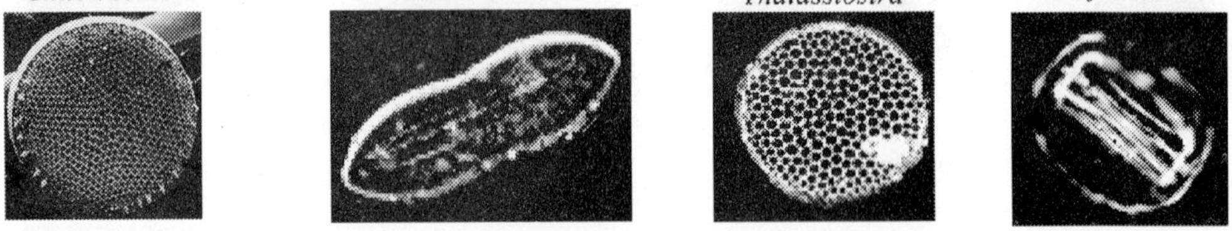

Thalassiosira *Psammodictyon* *Coscinodiscus* *Stellarima*

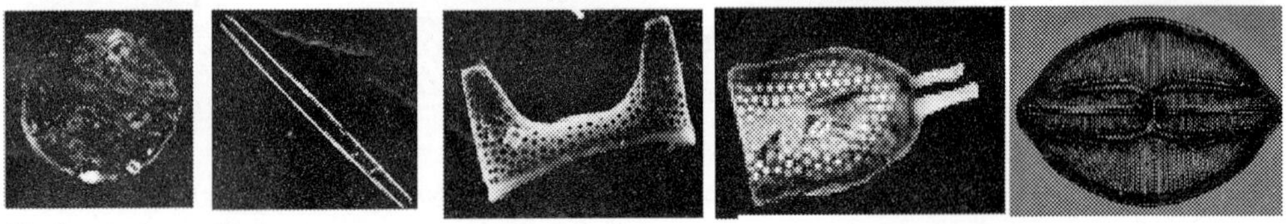

Thalassiosira *Thalassionema* *Eucampia* *Stephanopyxis* *Lyrella*

APPENDIX

Appendix - I

Proforma of Description for Foraminifera (Serial)

Phylum: Protozoa, Class: Sarcomastigophora
Order: Foraminiferida
Suborder: ....
Superfamily: ....
Family : ....
Genus: .... sp.

01	**Size of the test**	:	
02	**Shape**	:	
03	**Arrangement**	:	
04	**Width**	:	Gradually/Abruptly/Broadening at distal end
05	**Chambers**	:	1 in a plane / 2 in a plane / 3 in a plane
	(a) No. of Chambers	:	
	(b) Shape	:	Triangular/Quadrangular/Other
	(c) Enlargement	:	Gradual/Rapid
06	**Sutures**	:	Distinct/Indistinct – Raised/Depressed/ – Other special features
07	**Aperture**	:	Position – Tooth (Visible/Invisible) – Neck (Present/Absent)
08	**Periphery**		Entire/Lobulate Keeled/Without Keel
09	**Wall**		Agglutinated/Calcareous – perforate/Calcareous – imperforated
10	**Age**	:	
11	**Habitat**	:	Planktic/Benthic/Epifaunal/Infaunal
12	**Special Features**	:	

Lateral View

Apertural View

....

....

....

Proforma of Description for Foraminifera (Streptospiral, Miliolids)

Phylum: Protozoa, Class: Sarcomastigophora
Order: Foraminiferida
Suborder: ....
Superfamily: ....
Family : ....
Genus: .... sp.
Species........................... sp.

01 **Size of the test** : Large/Medium/Small

02 **Ratio of length to width:**

03 **Shape** : Elliptical/Sub-rounded/

04 **Arrangement of chambers** : Streptospiral
72° apart (Quinqueloculine added at 144°)/
120° apart (Triloculine)/180° apart (Spiroloculine)

05 **Aperture** : Neck (Present/Absent)
Lip (Present/Absent)

06 **Wall** : Pocellaneous/Agglutinated

07 **Surface ornamentation:**

(a) Reticulate/Pitted/Costate/
– Striae etc.

(b) Perforated/Imperforated

(c) Finely perforated/Coarsely
– perforated

08 **Number of chambers visible:**

09 **Age:**

10 **Habitat** : Benthic/other

11 **Special Features** :

....

....

....

Lateral View

Apertural View

Proforma of Description for Foraminifera (Coiled)

Phylum: Protozoa, Class: Sarcomastigophora
Order: Foraminiferida
Suborder: ....
Superfamily: ....
Family : ....
Genus: .... sp.
Species........................... sp.

01 **Size of the test** : Large/Medium/Small

02 **Shape of Test** : Biconvex/Plano-convex/Concavo-convex

03 **Arrangement of Chamber :**

(a) Shape of chambers (dorsally and ventrally)
Globular/Ovate/Clavate/Angular rhomboid/
- Angular-conical/Radial/Elongate

(b) Nature of final chamber:
Normal/Inflated/Deflated

(c) No. of cham. in the last whorl:

04 **Sutures (dorsally and ventrally) :**

(a) Appearance: Distinct/Indistinct
(b) Shape: Simple/Limbate
(c) Nature: Raised/Depressed/Flush
(d) Shape at periphery: Acute/Blunt
(e) Arrangement: Straight/Oblique
(f) Thickness: Uniform/Non-uniform

05 **Perforation :**

(a) Perforate/Imperforate
(b) Relative size of perforation: Finely/
- Coarsely perforate

06 **Features around umbilical area:**

(a) Plug: Present/Absent
(b) Size of plug: Large/.Medium/Small

07 **Periphery:**

(a) In lateral view: Entire/Lobulate
(b) In apertural view: Acute/Blunt/
- Rounded/Sub-rounded
(c) Keeled/Unkeeled

08 **Aperture:**

(a) Position: Terminal/Interiomarginal/
- Umbilical/at the base of apertual face
(b) Shape: Rounded/Slit like/Multiple pores
(c) Apertural lip : Present/Absent

Dorsal View

Ventral view

Apertural view

09. **Wall :** Calcareous/Agglutinated/Other

10 **Age :**

11. **Habitate:** Planktic/Benthic/Epifaunal/Infaunal

12 **Special Features**

....
....
....

Proforma of Description for Ostracoda

(Superphylum : Anthropoda, Phylum : Crustacea, Class : Ostracoda)

Order : ....
Suborder: ....
Superfamily: ....
Family : ....
Genus: ....sp.
Species..................... sp.

01. **Basic shape of Carapace:**
Elongate/Ovate/Sub-rectangular/Sub-cylindrical
Sub-triangular/Semicircular/Sub-trapezoid

02. **Nature of Carapace:**
Normal / Compressed/Inflated

03. **Ratio of length to width :** 1 :

04. **Boundaries:**
   (a) Dorsal: Straight/nearly straight/ -Concave/convex
   (b) Ventral: Straight/nearly straight/ - Concave/convex

05 **Ends :**
   (a) Anterior
   (b) Posterior

External View

06 **Surface :**
   (a) Smooth
   (b) Ornamented
      (i) Weakly/Strongly
      (ii) Type of ornamentation: Reticulation/Spines/Tubercles/Other

07 **Eye Spot** : Present/Absent

08 **Muscle Scars** Position : Central / other

Internal View

09 **Hinge** : Adont/Merodont/Entomodont/Amphidont

10 **Age** :

11 **Habitat** : Planktic/Benthic/other

12 **Dimorphism** :

13 **Special Features :**

....

....

....

References

Ahmed, S. T., 1995. Calcareous Micropaleontology of the Neogene of Cox Bazar, Bangladesh. *Rev. Espanola Micropal.,* XXVI, **1**, 31-68.

Aldrige, R. J., Briggs, D. E. G., Clarkson, E. N. K. and. Clark, N. D. L., 1983. The Conodont Animal. *Lethaia*, **16**, 1-14.

Anderson, D. M., Overpeck, J. T. and Gupta, A. K., 2002. Increase in the Asian southwest monsoon during the past four centuries. *Sci.*, **297**, 596-599.

Angel, M. V., 1969. Planktonic ostracods from the Canary Island region: their death distribution, diurnal migration, and community organization. *J. Mar. Bio. Assoc.,* U. K., **49**, 515-33.

Armstrong, H. A. and Brasier, M. D., 2005. **Microfossils.** Blackwell Publishers, 296 p.

Awerinzew, S. O., 1901. Struktur der kalkschalen mariner Rhizopoden: Sankt-Petersb. *O-vo Estesv.,* Turdy, **32**.

Banner, F. T., 1971. A synopsis of the Spirocyclanidae. *Rev. Espan. Micropal.,* **2**, 243-290.

Bate, R. H. and East, B. A.,1972. The structure of ostracode carapace. *Lethia,* **6**, 177-94.

Beccarius, J. B., 1731. De bononiensi arena quadam. *Com. Bonon. Sci. Art. Inst.,* **1**.

Benson, R. H. , 1961.Ostracoda, In Treatise on Invertebrate Paleontology, Pt. Q, Arthropoda 3, Crustacea. R. C. Moore (ed.), *Geol. Soc. Am. & Univ. Kansas Press.*

Berger, W. H. and Gardner, J. V., 1975. On the determination of Pleistocene temperatures from planktonic foraminifera. *J. Foram. Res.,* **5**, 102.

Berggren, W. A., 1972. A Cenozoic time-scale--Some implications for regional geology and paleobiogeography: Lethia, **5** (2), 195-215.

Bernard, A., 1995. Trachyleberididae (Ostracodes) du Turonion Superier (?)- Santonien de la region de Boulrnane, Moyen Atlas (Maroc): Systematique et biostratigraphie. *Rev. Espanola Micropal.,* XXVII, **1**, 85-142.

Bhalla, S. N. and Kathal, P. K., 1998. Recent foraminiferal thanatocoenose from Gulf of Mannar. *N. Jb. Geol. Pal., Abh.*, Germany, 207 (3), 419-431.

Bhandari, A., 1995. Ostracoda from Intertrappean Beds near Duddukuru, Andhra Pradesh and a note on their age, paleoecological significance. *Ind. J. Petrol. Geol.,* **4**(1), 89-107.

Bolli, H.M. and Saunders, J.B., 1985. Correlation with the deep sea and other planktonic fossil records, Oligocene to Holocene low latitude and Cenozoic foraminiferal and dinoflagellate cyst bistratigraphy of the central North America.

Boltovskoy, E. and Wright, R., 1976. **Recent Foraminifera**. The Hague: Dr. W. Junk, b v.

Boltovskoy, E., Watande, S., Totah, V. and Vera Ocampo, J., 1995. Benthic foraminifera from DSDP Site 516 (Upper Maestrichtian-Quaternary, South Atlantic). *Rev. Espanola Micropal.,* XXVII, **2**, 111-141.

Brady, H. B., 1884. Report on the Foraminifera dredged by H. M. S. Challenger, during the years 1873-1876. *Challenger Rep. Zool.*, **9**(22), 21-814, pl. 115.

Bramlette, M. N. and Sullivan, F. R., 1961. Coccolithophorids and related nannoplankton of the early Tertiary in California. *Micropal.*, **7**, 129-188.

Bramlette, M. N. and Wilcoxon, J. A., 1967. Middle tertiary calcareous nannoplankton of the Cipero section, Trinidad, W. I. *Tulane Studies in Geology and Paleontology*, **5**, 93-131.

Bramlette, M. N. and Riedel, W. R., 1954. Stratigraphic value of discoaster and some other microfossils related to Recent coccolithophores. *J. Pal.*, **28**, 385-403.

Brasier, M. D., 1980. **Microfossils**. George Allen & Unwin, 193 p.

Brasier, M. D., 1977. An early Cambrian chert biota and its implications. *Nature*, **268**, 719-720.

Brasier, M. D., 1986. Why do lower plants and animals biomineralize? *Paleobiology*, **12**, 241-50.

Briggs, D. E. G., Clarkson, E. N. K. and Aldrige, R. J., 1983. The Conodont Animal. *Lethaia*, **16**, 1-14.

Burkey, D. 1971. Cenozoic calcareous nannofossils from the Pacific Ocean. *Trans. San Diego So at. Hist.*, **16**, 303-328.

Carpenter, W. B., 1856. **Researches on the Foraminifera**. In: *Roy. Soc., Lond.*

Carpenter, W. B., Parker, W. K. and Jones, T.R. 1862. **Introduction to the study of Foraminifera**. *Roy. Soc. Lond.*,319 p.

Chapman, F. and Parr, W. J. 1936. **A classification of Foraminifera**. *Roy. Soc. Victoria, Proc.*, **49**(1).

Colom, G., 1956. Lito-facies y micropaleontologia de las formaciones jurasico-neocomienses de la sierra de Ricote (Murcia). *Bull. Intst. Geol. Min. Spain,* Madrid, **67**, 11-67.

Conway-Morris, S., 1976. A new Cambrian lophophorate from the Burgess Shale of British Columbia. *Pal.*, **19**, 199-222.

Cooper, S. C., 1964. Benthonic Foraminifera of the Chukehi Sea, *Cush. Lab. Foram. Res. Contr.*, **15**(3), 89-104.

Coryell, H. N. and Fields, S., 1937. A Gatun ostracode fauna from Cativa, Panama. *Am. Mus., Novitates,* **956**, 1-18.

Cushman, J. A., 1925. An introduction to the morphology and classification of the Foraminifera. *Smithsonian Misc.. Coll.*, **77**(4), 1-77, pls.16.

Cushman, J. A., 1927. An outline of a re-classification of the foraminifera: *Cush Lab. Foram. Res. Contr.*, **3**(1), 1-105.

Cushman, J. A., 1948. **Foraminifera their classification and economic use**. 4th ed., Cambridge, Mass, Har. Univ.Press.

d' Orbigny, Alicide, 1826. Tableau methodique de la classe des Cephalopodes, 3 me ordre Foraminiferes, *Ann.Sci. Nat.*, **7**, 245-314.

d'Orbigny, Alicide, 1839. Foraminiferes. In: Raman de la sagra, *Hostoire, Politique et naturelle de I'llede Caba*. Paris, **8**, 224p., 22pls.

d'Orbigny, Alicide, 1846. Foraminiferes, Fossiles Du Basin Tertiare De Vienne. Gide at caop., Libraries-editeur No. 5.

De Blainville, H. M. D. De and Von Wolderheim, 1834. Manuel d'Actinologie ou de Zoophytologie. (F.G. Levrault: Paris, Strasbourg): **1** (text), 1-viii, 1-644, **2** (atlas), pls I-C.

Dragastan, O., Golubic, S. and Richeter, D. K., 1996. Rivularia Haematites: a case of Recent verses fossil morphology. Taxonomical considerations. *Rev. Espanola Micropal.,* XXVII, **1**,42.

Dujardin, F., 1835. Observations nouvelles sur less pretendus Cephalopods microscopiques. *Ann. Sc. Nat.*, **2**(3), Zool., 36p.

Dwivedi, P., Chandra, U., Dhawan, R., Mittal, A., Unival, A. K. and Tripathi, G. K., 1995. Stable carbon isotope geochemistry of associated gases of Gandhar field, Cambay basin, India. *Ind. J. Petro. Geol.,* **3**(1), 64-79.

Edwards, A. R., 1971. A calcareous nannoplankton zonation of the New England Paleogene, *Proc. 11 Plank. Conf.,* Rome, 381-419.

Ehrenberg, C.G. 1854. Mikrogeologie. Einundvierzig Tafeln mit über viertausend grossentheils colorirten Figuren, Voss, Leipzig., Pl. 5, II, fig. 23 (iconotype).

Eicher, D. L., 1969. Paleobathymetery of Cretaceous Greenhorn sea in Eastern Colorado, *Am. Assoc. Petrol. Geol.*, **53**, 1079.

Ellis B. F. and Massina, A. R., 1940. to date **Catalogue of Foraminifera**, spec. pub., *Am. Mus. Nat. Hist.* over 50 volumes (microfilm).

Emiliani, C., 1955. Pleistocene temperatures. *Jour. Geol.,* **63**, 538-578.

Fay, R. O., 1952. **Catalogue of conodonts**. Univ. Kansas. Paleo. Contr., Vertebrata, Art. **3**, 1-206, 1-109 figs.

Frerichs, W. E., Heiman, M. E. Borgmen, L. E., and Be, A. W. H., 1972. Latitudinal variation in planktonic test porosity. *J. Foram. Res.,* **2**, 6-13.

Funnell, B. M., 1967. Foraminifera and Radiolaria as depth indicators in the marine environment. *Mar. Geol.,* **5**, 333-347.

Funnell, B. M. and Riedel, W. R. (eds.)., 1971. **The Micropaleontology of oceans**. Cambridge Univ. Press, 311p.

Galloway, J. J., 1933. **A manual of Foraminifera, Indiana**. Bloomington, Principia Press.

Gartner, S., 1969. Phylogenetic lineages in the Lower Tertiary coccolith genus *Chaismolithus, Proc. N. Am. Pal. Conv.,* G, 930-957.

Gibson, T. G., 1989. Miocene evolution of *Tubulogenerina* in the Indo-Pacific and african areas. *J. Foram. Res.,* **19**(2), 126-145.

Glaessner, M. F., 1945. **Principles of Micropalaeontology**. Melbourne University Press, 296p.

Goldstein, S. T., 1999. Foraminifera: a biological overview, *In*: **Modern Foraminifera,** Sen Gupta, B K., Kluwer Academic Publisher, 37-56 (Eds.).

Guha, D. K., 1986. The study of Ostracodes, how important they are in oil exploration, *In*: *Proc. Ind. Colloq. Micropal., Stragr., 579-584.*

Gupta, A. K., Das, M., Clemens, S. C. and Mukherjee, B., 2008. Benthic foraminifera faunal and isotope changes as recoreded in Holocene sediments of the north west Indian Ocean. *Paleooceanogr.*, **23**, PA2214, 1-10.

Gupta, A. K., Karas, C., Nurnburg, D., Tiedemann, R., Mohan, K. and Bickert, T., 2009. Mid-Pliocene climate change amplified by a switch in Indonesian subsurface throughflow. *Nature Geosci.,* **2**, 434-438.

Gupta, A. K., Singh, R. K., Joseph, S. and Thomas, E., 2004. Indian Ocean high productivity event (10-8 Ma): linked to global cooling or to the initiation of the Indian Monsoons? *J. Geol. Soc. Am.*, **32**(9), 253-256.

Hanjo, S. and Okada, A.B., 1974. **Oceanographic Micropalaeontology**. **2**, London: Academic Press.

Hanjo, S., 1976. Coccoliths: production transportation, and sedimentation, *Mar. Micropal.,* **1**, 65-79.

Hansen, H. J., 1968. X-ray diffractometer investigations of a radiate and granulate Foraminifera, *Dan. Geol. Foren., Medd.,* **18**, 346-348.

Hansen, H. J., 1972. Two species of Foraminifera of *Turrilina* with different wall structure. *Lethia,* **5**, 39-45.

Haq, B. U. and Van Couvering, J. A., 1977. Corrected age of Pliocene/Pleistocene boundary, *Nature,* **269**, 243-288.

Haq, B. U., 1978. Calcareous nannoplankton, **Introduction to marine Micropaleontology**. *In*: B. U. Haq and Boersma (eds.), Elsevier 78-107.

Hasle, G. R. and Syvertsen, E. E., 1997. Marine diatoms. In Tomas, C. R. (ed.), Identifying Marine Phytoplankton. Academic Press, San Diego, 5–385.

Hass, W. H., Hantzschel, W., Fisher, D. W., Howell, B. H., Rhodes, F. H. T., Muller, K. J. and Moore, R. C.,1963/64. Misscellanea. In: *Tretise on Inver. Pal.,* pt. W. R. C. Moore (ed.) 3-98. *Geol. Soc. Am., Kansas Univ. Press.*

Hendy, N. I., 1964. An introductory account of smaller algae of British Coastal Water, Part; V. Bacillariophyceae (Diatoms) Min. Ag. Fish. Food Fishery

Hillock, P., 1999. **Modern Foraminifera**. *In*: (ed.) Sen Gupta, B. K., Kluwer Academic Publishers123-129.

Hilterman, H., 1949. **Classification der naturlichen Brackwasser**. Erdolu. Koble, **2**, 4-8.

Hodell, D. A, McKenzie, L. A., Mueller, P. A and Muller, D. A., 1987. Application of the strontium isotope sea water curve as a chronometer for delimiting late Miocene/early Pliocene event stratigraphy, In: *Fourth Int. Cong. Pacific Neogene Stratigr.,* Abst. p. 39.

Hofker, J., 1953. Arenaceous tests in foraminifera chalk or silica (?). *The Micropal.*, **7**(3), 65-66.

Holme, N. A. and McIntryre, A. D., 1984. **Methods for the study of marine benthos. IBP Handbook** 16, 2nd Edition, Blackwell Scientific Publications, Oxford, 24op.

Hottinger, L., 1982. Larger foraminifera, giant cells with historical background. *Naturwissenschaften*, **69**, 361-371.

Hustedt, F., 1930. **Bacillariophyta (diatoms)**. In: *Die Suswasserflora Mitteleuropas*, A. Pascher (ed.). Gena G. Fishcher, 332p.

Huxley, 1868. **On a piece of chalk.** Re-printed 1967, Scribner's New York, 90p.

Jones, D. J., 1956. **Introduction to Microfossils**. Harper & Brothers, New York, 405p.

Kathal, P. K., 1989. **A study on Recent Foraminifera from the sandy beaches of east coast of India** (Ph.D. thesis,, Aligrah), unpublished.

Kathal, P. K., 1996. Foraminiferal ecophenotypes in paralic environment: Significance in paleoenvironmental interpretations, In: *Proc. Ind. Geol. Cong.,* 15-19.

Kathal, P. K., 1998. **Microfossils and their applications**. CBS Pub., New Delhi, 145p.

Kathal, P. K., 1999. Foramgeographical affinity of Indian Ocean during Quaternary based on the Recent foraminiferal thanatocoenoses, Gulf of Mannar, India. *Gond. Geol. Mag*., sp. **4**, 281-292.

Kathal, P. K., 2002a. Taxonomy, distribution pattern and ecology of the Recent littoral foraminifera of the east coast of India. *N. Jh. Geol. Pal., Abh.*, (Germany), **224**(1), 115-160.

Kathal, P. K., 2002b. Distribution and ecology of Recent foraminifera from littoral sediments of eastern India. *J. Geo. Soc. India,* **60**, 429-454.

Kathal, P. K., 2005. Sethusamudram Ship Canal Project: Oceanographical/Geological and Ecological Impact on Marine Life of Gulf of Mannar and Palk Bay, Southeastern Coast of India. *Curr. Sci.*, **89** (7), 1082-1083.

Kathal, P. K., 2008. Foramgeographical affinity of the Indian Coasts: an overview based on statistical studies. *In*: 5th *Inter. Conf. Environ. Micropal. Microbio. Meioben.*, (ISEMMM 2008), Univ. of Madras, 143-145.

Kathal, P. K., 2011. Morphological variations in common recent benthic foraminifera from the east coast of India and the southern east coast of Japan, *J. Pal. Soc. Ind.*, **56**(1), 65-81 (jointly with Singh, V. K).

Kathal, P. K. and Bhalla, S. N., 1996a. On the first report from the Indo-Pacific region, migratory trends and paleolatitudinal significance of *Rotorboides granulosum*- a tropical water Foraminifera since Middle Pliocene (Late Miocene?), *Contr. XV Indian Colloq. Micropal, Stratgr., 317-320.*

Kathal, P. K. and Bhalla, S. N. 1996b. Intraspecific variation and paleolatitudinal significance of *Rotorboides granulosum* - less known Recent Foraminifera of tropical waters. *Rev. Paleobio.* (Switzerland), **15**(1), 79-95.

Kathal, P. K and Bhalla, S. N. 1998. Taxonomy and paleolatitudinal significance of *Rotorboides granulosum*- a less known Foraminifera from tropical waters. *J. Geol. Soc. India*, **51**(6), 799-802.

Kathal, P. K. and Bhalla, S. N. 2001. Blending of Recent foraminiferal biogeographical provinces of India Waters- a statistical approach. *N. Jb. Geol. Pal.,* (Germany), **2001**(4), 250-256.

Kathal, P. K., Bhalla, S. N. and Nigam, R. 2000. Foramgeographical affinities of the west and east coasts of India: An approach through cluster analysis and comparison of taxono-mical, environmental and ecological parameters of Recent foraminiferal thanatotopes. *Bull. Oil & Nat. Gas Corp., India,* **37**(2), 65-75.

Kathal, P. K. and Singh, V. K., 2010. First report of some Recent benthic foraminifera from the east coast of India. *J. Geo. Soc. India,* **76** (1), 69-74.

Kathal, P. K. and Singh, V. K. Comparative morphology of the common benthic Recent foraminifera from the east coast of India and Japan and the foramgeographical affinity of the two regions of the Indo-Pacific realm. *J. Pal. Res.* (Japan), (*in press*).

Kathal, P. K. and Singh, V. K. Biogeographical affinity in the Indian and Japanese waters based on the Recent foraminifera, east coast of India and east coast of Japan *Curr. Sci.*, (In press).

Kathal, P. K. and Singh, V. K. Revisit to K/T Boundary, Jhilmili Section and detailed survey of Intertrappens sediments, Chhindwara District, M. P. (*editorial hands, Curr. Sci.*).

Keller, G. 1985. Depth stratification of planktonic foraminifers in the Miocene Ocean, *Geol. Soc. Am. Mem.* **163**, 177-195.

Keller, G., Sahni, A. and Bajpai, S. (2009) Deccan Volcanism, the KT mass extinction and dinosaurs. *J. Biosci*, **34**(5), Ind. Acd. Sci., 709-728

Kennett, J. P. and Stott, L. D. 1991. Abrupt deep-sea warming paleooceanographic changes and benthic extinction at the end of Paleocene. *Nature*, **353**, 225-229.

Kesling, R. V. 1951. Mechanical solution of formulas for growth rates. *Contr. Mus. Paleo. Univ. Mich.*, **8**(10), 231-237.

Khosla, S.C. 1978. Lower Miocene Ostracoda from Jamnagar and Porbandar District, Gujrat, India. *Micropal.*, **24**(3), 251-290.

Khosla, S. C., Nagori, M. L., Jhakar, S. R. and Rathor, A. S. (2009). Mixed marine, brackish water and non-marine microfaunal association in the Inter-trappen Beds (Early Plaeocene) of Jhilmili, Chhindwara, Madhya Pradesh, *J. Geol . Soc. Ind.*, **73**, 724-732.

Krashennikov, V. A. 1960. Elfidiidy miotsenovykh otlozheniy podolii (Elphidiidae of the Miocene strata of Odolia). *Trudy geologiecheskogo instituta, Akademiya Nauk SSSR*, Moscow, **21**, 1-141.

Kureshy, A. A. 1958. The paleoecology of Foraminifera, *In*: *XXII Inter. Geol. Cong., Proc., 257-266.*

Lamarck, J. B. 1801. **Systeme des animaux sans vrtebres**. Paris. The Author, 432p.

Lamarck, J. B. 1822. **Historie naturelle des animux vrtebres**. Paris L'auteur, **7**, 711p.

Landing, E. 1977. *Prooniotodus tenus* (Muller, 1959) apparatus from the Taconic allocthon, eastern New York: Construction taphonomy and the protoconodont 'supertooth' model. *J. Pal.*, **51**, 1072-1084

Lee, J. J. 1974. Towards understanding the niche of foraminifera. *In*: Hadley, R.H. Adams, C.G. (eds.), *Foraminifera*, **1**, 207-260.

Levine, N. D. 1962. Protozoology today, *J. Protozool.*, **9**(1), 1-6.

Linne, C. 1758. Systema natura per reganatria nature secundum classes, oredines genera, species kcum characterlibus differentis synonumis locis, **10**(1), 824p., Engelman, (Lipsiae).

Loeblich and Tappan, H. 1966. Annoted index and bibliography of the calcareous nannoplankton. **I.** *Phycologia,* 5, 81-216; **II.** 1968. *J Pal.,* 42, 584-598; **III**. 1969. *J Pal.,* 43, 568-588; **Iv.** 1970. *J. Pal.,* 44, 558-574; **V.** 1970, *Phycologia;* 9, 157-174; **VI.** 1971. *Phycologia,* 10,315-339; **VII.** 1973. *J Pal.,* 47, 715-759.

Loeblich, Jr. A. R. and Tappan, H. 1988. **Framiniferal genera and their classification**. Van Nostrand Reinhold Company. New York, 970p., 874 pls.

Loeblich, Jr. A. R. and Tappan, H. 1964. Protista 2: Sarcodina, chiefly "Thecamoebians' and Foraminiferida. In **Treatise on Invertebrate Palaeontology**, Part C, R. C. Moore, (ed.), (2 vols). *GeoI. Soc. Am., Univ. Kansas Press.*. U.S.A.

Loeblich, Jr. A. R. and Tappan, H. 1984. Supragreneric classification of Foraminiferida (Protozoa). *Micropal.*, **30**(1). 1-69.

Lord, A. R. and Taylor, R. J. 1982. **A Stratigraphic Index of Calcareous Nannofossils** (eds. Lord, A.R. and Taylor, R.J.). Ellis Harwood Ltd., U. K., **9**.

Lynch-Stieglitz, L. and Fairbanks, R G. 1993. Inter-ocean exchange of glacial North Atlantic Intermediate waters., *Nature*, **369**, 308-310.

Marszalek, D. S., Wright, R. C. and Hay, W. W. 1969. Function of the test in the foraminifera. *Trans. Gulf. Coast. Ass. Geol.. Socs.*, **19**, 341-352.

Martini, E., 1971. Standard tertiary and Quaternary Calcareous nannoplankton zonation. *Proc. II Plank., Cong.*, Rome, 739-785.

McGlasson, R. H. l959. Foraminiferal biofacies around Santa Catalina Island, California, *Micropal.*, **5**, 217-240.

McIntyre, A. 1967. Coccoliths as Paleoclimate Indicators of Pleistocene Glaciation, *Sci.*, **158** (3806), 1314-1317.

Mckenzie, K. G. 1982. Homoeomorphy: Persistent joker in the taxonomic pack, with the description of *Bradleycypris* gen. nov., **Fossil and Recent Ostracods** (eds. Bate, R .H., Robinson, E. & Shepperd, L. M.) *In*: British Micropal. Soc., Ellis Horwood Sr., Geol., 407-438.

Meglitsch, P. A. and Schram, F. R. 1991. **Invertebrate Zoology**, (III ed.), Oxford University Press.

Miller, D. N. 1953. Ecological studies of Foraminifera of Mason Inlet, North Carolina, *Cush. Found. Foram. Res. Contr.*, **4**(2), 41-63.

Monfort, P. D. 1808. Conchiologie Sistematique et classification Methodique des Coquilles, Paris, 409.

Montegu, G. 1803-1808. Testacea Britanica, or Natural History of British Shells, Marine Land and Freshwaters, London. 3, 606 p.

Murray, J. W. 1973. Distribution and ecology of living benthic foraminiferids, London, Heinemann.

Murray, J. W. 1991. **Ecology and paleoecology of benthic foraminifera**, Longman Scientific and Technical, New York. 380.

Myers, A. H. 1943. Life activities of Foraminifera in relation to marine ecology, *Am. Phil. Soc., Proc.*, **86**(3), 439-458.

Nagappa, Y. 1957. Direction of coiling in *Globorotalia* as an aid in correlation. *Micropal.*, **3**(4), 393-398.

Nigam, R. and Khare, N. 1994. Signals of the change in monsoonal precipitation at around 2,000 year BP in a sediment core, off central west coast of India. *Curr. Sci.* **66**, 226-228.

Nigam, R. 2005. Addressing environmental issues through foraminifera– Case studies from the Arabian Sea. *J. Pal. Soc. Ind.*, **50**(2), 25-36.

Nigam, R., Khare, N. 1999. Spatial and temporal distribution of foraminifera in sediments off the central west coast of India and use of their test morphologies for the reconstruction of paleomonsoonal precipitation. *Micropal.* **45**(3), 285-303.

Nigam, R., Chaturvedi, S. K. 2006. Do inverted depositional sequences and allochthonous foraminifers in sediments along the Coast of Kachchh, NW India, indicate palaeostorm and/or tsunami effects? *Geo-Mar. Lett.*, **26**, 42-50.

Nigam, R., Panchang, R., Banerjee, P. 2005, Foraminifera in surface sediments of Mandovi River Estuary: Indicators for mining pollution and high sea stand in Goa, India. *J. Coast. Res.*, **21**(4), 853-859.

Nigam, R., Prasad, V., Saraswat, R., Garg, R., Mazumder, A. and Henriques, P.J. 2009. Late Holocene changes in hypoxia off the west coast of India: Micropaleontological evidences. *Curr. Sci.*, **96**, 708-713.

Nigam, R., Saraswat, R., Kurtarkar, S.R. 2006. Laboratory experiment to study the effect of salinity variations on benthic foraminiferal species - *Pararotalia nipponica* (Asano). *J. Geol. Soc. Ind.*,**67**(1), 41-46.

Norton, R. D. 1930. Ecologic relation of some Foraminifera, *Bull., Sripps Inst. Oceangr. Tech.,* **2**(9), 331-338.

Oretli, H. J. 1971. The aspects of Ostracod faunas. A possible new tool in Petroleum Sedimentology. *Bull. Cen. Res.* Paris, SNPA**,** 137-151.

Panchang, R., Nigam, R., Baig, N., Nayak, G.N. 2005. A foraminiferal testimony for the reduced adverse effects of mining in Zuari Estuary, Goa. *Int. J. Environ. Stud.*, **62**(5), 579-591.

Panchang, R., Nigam, R., Linshy, V.N., Rana, S.S., Ingole, B.S., 2006, Effect of oxygen manipulations on benthic foraminifera: A preliminary experiment. *Ind. J. Mar. Sci.*, **35**(3), 235-239.

Panchang, R., Nigam, R., Raviprasad, G.V., Rajagopalan, G., Ray, D.K., Hla, U Ko Yi, 2008. Relict faunal testimony for sea-level fluctuations off Myanmar (Burma). *J. Pal. Soc. Ind.,* **53**(2), 185-195.

Pander, C.H. 1856. Monographie der fossilen Fische des silurichen Systems der russisch-baltischen Gouvernments. K. Akad. D. Wiss., St. Petesburg, 1-91, 1-9pls.

Pele-Alempy, A.M. and Wuchang, Wei 1995. Megnetobiochronology of several Miocene nannofossil datums at DSDP Site 563, Revisited. *Rev. Espanola Micropal.*, XXVII, **2**, 97-110.

Phleger, F.B. 1960a. Foraminifera population in Laguna Madre, Texas coast. *Cush. Found. Foram. Res. Contr.* **7**(4), 106-151.

Prins, B. 1960. **Ecology and distribution of Recent Foraminifera**, John Hopkms Press, Bahimore, 297 p.

Prins, B. 1969. Evolution and stratigraphy of coccolithinids from the Lower and Middle Lias. *Proc. I, Plank. Conf.*, Geneva, **2**, 547-558.

Ramane, A. 1934. Die Bracbwasser-Fauna (mit besonderer Burucksichitigung der ostee). *Verhandl deuschen. Zool, Ges,* **36**, 34-74.

Rauzer-Chernousova, D.M. and Fursenko, A.V. 1959. Fundamentals of paleontology Osnovy paleontologii): A manual for paleontologists and geologists of USSR, gen. pt. Protozoa, Yu. A. Orlov, (chief ed.), *Izdatel'stvo Akad. Nauk SSR,* Moscow.

Reiss, Z. 1963. Reclassification of perforate Foraminifera. *Geol. Surv.Israel, Bull.,* **35**. 1-111.

Reuss, A. 1862. Entwurf einer systematisschen Zusammenstellung der Foraminiferen, *Akad, Wiss, Wien, S.-B., Math, Nat. Kl.,* Abt., **1**(44), 355-396.

Rhumber, L. 1895. Entwuf eines naturleuchen Systems der Thalamophoren, *Ges. Wiss. Gottingen, Math. Phys. Kl. Nachr.* **1**,51-98.

Roth, P. H. and Berger, W. H. 1975. Distribution and dissolution of coccoliths in the south and central pacific. In: Dissolution of Deep sea Carbonates. *Cush. Found. Foram Res. Contr.,* **13**, 87-113.

Round, F. E., Crawford, R.M. and Mann, D.G. 1990. The Diatoms: biology & morphology of the genera. Cambridge University Press, 747 p.

Said, R. 1950. the distribution of Foraminifera in the northern Red Sea, *Cush. Found. Foram. Res. Contr.,* **1**(1), 9-29.

Saidova, Kh. M., 1961. Ehkologia foraminifer i paleogeografia dal'neostocbnykb. morej SSSR I severozapadanoj chasti Tikhogo, *Akad. Nauk SSSR, Inst. Okenol,* 182 p.

Saidova, Kh. M., 1967. Sediment stratigraphy and paleogeography of the Pacific Ocean by benthonic Foraminifera during the Quaternary, *Progress in Oceanogr.,* **4**, 143-151.

Saraswat, R., Nigam, R. and Barreto, L. 2005. Paleoceanographic implications of abundance and mean proloculus diameter of benthic foraminiferal species *Epistominella exigua* in sub-surface sediments from distal Bay of Bengal Fan. *J. Earth Sys. Sci.,* **114**, 453-458.

Saraswat, R., Nigam, R. and Pachkhande, 2011. Difference in optimum temperature for growth and reproduction in benthic foraminifer *Rosalina globularis*: Implications for paleoclimatic studies. *J. Experimental Mar. Bio. Eco.,* **405**, 105-110.

Savin, S. M. and Douglas, R. G. 1973. Stable isotope and magnesium geochemistry of Recent Planktonic Foraminifera from the South Pacific, *Geol. Soc. Amer., Bull.,* **84**, 2327-2342.

Schiller, J., 1930. **Coccolitbinae**, in Robenhorst. (ed.) Krptogamane-flora von Deutschland, *Osterreich Schweiz,* **10**, II, 89-267.

Schmidt, H. 1934. Conodonten-funde in ursprungglichen Zusammenhang. *Pal. Zeischr.,* **16**, 76-85.

Schultze, M. S. 1854. Uber den Organismus der Polythelamien (Foraminiferen) nebst Bemerkungen uber die Rhizopoden im Allgemeinen, Engalmann. *Leipzig,* 68 p.

Scott, G. H., 1991. Basal Miocene correlation: *Globigerinoides* from the Caribbean and New Zealand. New *Zea., Micropal.,* **16**, 385-398.

Scott, H. W. and Smith, W.H. 1951. Molt stage of an Eocene fresh-water ostracode. *J. Pal.,* **25**(3), 327.

Scott, H. W. 1934. The zoological relationship of the conodonts, *J. Pal.,* **8**, 448-455.

Seibold, I., 1971. *Ammonia* Brunnich (Foram.) und Varwandate Arten Indischen Ozean (Malabar-Kuste, SW Indien). *Pal., Z.,* **45**, 41-52.

Selvester-Bradley, P. C. 1956. The structure, evolution and nomenclature of the ostracoda hinge. *Bull. Brit. Museum Geol.,* **3**(1), 1-21.

Serova, M. Ja. 1961. Vid *Trochammina vitrea* Serova sp.n., ego paleontologia; stratigraphischeskoe machenie. *Vopr. Micropal.,* **5**, 79-82.

Setty, M.G.A.P. 1979. Role of foraminifera in oceanographic events. *J. Sci. Indus. Res., Ind.*, **38**, 380- 399.

Shackelton, N.J. and Kennett, J.P. 1975. Palaeotemperature history of the Cenozoic and the initation of Antarctic glaciation: oxygen and carbon isotope analysis in DSDP sites 277, 279 & 281. *In: Initial Rep.*, DSDP, **29**, 743-756.

Sigal, J. 1952. Order des foraminifera. In Piveteau, *Triate de Pa.*, Masson, Paris, 133-301.

Singh, V.K. 2009, Study of comparative morphology of common Recent benthic foraminifera from the east coast off Tiba province, east coast of Japan (un published) Dr. Harisingh Gour University, 173p.

Singh, V. K. and Kathal, P. K.. (2010). First report of twelve benthic foraminifera from Indian waters. Gond. Geol. Mag., **25**(c), 259-266.

Singh, V.K. and Kathal, P.K. 2011. Morphological variations in common recent benthic foraminifera from the east coast of India and the southern east coast of Japan, *J. Pal. Soc. Ind.*, **56**(1), 65-81.

Soldani, A. 1789. Testaceografia et Zoophytografia, *Parva microsopica*, **1**, Siena, 1-80.

Sorby, H.C. 1861. On the organic origin of so-called `Crystalloides` of the Chalk, *Annals Mag. Nat Hist.*

Sorby, H.C. 1897. Anniversary address of the President, *Gool. Soc. London*, Proc., 1878-79; *Geol. Soc. Lon., Q.J.,* **35**, *Appendix* 56-93

Srinivasan, M.S. 1996. Oceanic Micropaleontology: major advances since the beginnjng of DSDP, Presidential Add., In: *XV Ind Colloq. Micropal. Strat.,* 34 p.

Sujata K.R., Nigam, R., Saraswat, R. and Linshy, V.N. 2011. Regeneration and abnormality in benthic foraminifera *Rosalina leei*: Implications in reconstructing past salinity changes. *Rivista Ital. Pal. Stratigr.* **117**, 189-196.

Swain, F.M., Petters, S.W. and Chuanti, Xie. 1996. Some Cretaceous ostracoda from Nigeria, *Rev. Esp. Micropal.,* XXVI, **2**, 1-5.

Thierstein, H.R. 1973. Lower Cretaceous calcareous nannoplankton biostratigraphy. *Abh Geol. Bundesnasi,* (Wim), **29,** 52.

Thiertstein, H.R. 1974. Calcareous nannoplankton- Leg 26 of the Deep Sea Drilling Project. In: *Initial Reports D.S.D.P., XXXVl.,* U. S. Government Printing Office, Washington, D. C., 619-667.

Thiertstein, H. R., 1976. Mesozoic calcareous nannoplankton biostratigraphy of marine sediments. Mar. *Micropal.*,**1**, 325-362.

Todd, R. and Blackmon, P. 1956. Calcite and aragonite m Foraminifera. *J. Pal.,* **30**, 217-219.

Towe, K.M. and Cifelli, R. 1967. Wall structure m the calcareous Foraminifera, crystallographic aspects and the model for calcification. *J. Pal..,* **41**(3), 742-762.

Trejo, M.H. 1960. La Famila Nannoconidae y su alcance estratigrafico en America (protozoa, Incertae saedis). *Bull. Assoc. Mex. Geol. Petro.*, XII, 259-314.

Triebels, E. 1941. Zur Morphologie und Oekologie der fossilen Ostracoden. Mit Bescheibuhg emiger neuer Gattagen und Arten. *Senkenbergiana,* 23 4-6, 294-400.

Ulrich, E.O. and Bassler, R.S. 1926. A classification of the tooth-like fossils, conodonts, with descriptions of American Devonian and Mississippian species, *Proc. U.S. Nat. Mus.* **68**, art. 12, 2613, 1-63, 11pls., 1-5 figs.

Ulrich, E.O. and Bassler, R.S. 1926. A classification of the tooth-like fossils, conodonts, with descriptions of American Devonian and Mississippian species. *Proc. U.S. Nat. Mus.*, **68**, Art. 12, 1213, 1-63, 11 pls., figs. 5.

Urey, H.C. 1947. The thermodynamic properties of isotopic substances. *J. Chem. Soc.*, **1**, 562-581.

Van der Bold, W.A. 1946. **Contribution to the study of Ostracoda**. *Amsterdam, J. H., De Bussy*, 167 p.

Van Hinte, J.E., 1978. Geohistory analysis- application of micropaleontology in exploration geology. *AAPG Bulletin*; **62**(2), 201-222.

Van Markhovan, F.P.C.M., 1962. Post Paleozoic Ostracoda, their mineralogy, taxonomy, andeconomic use. 1. Elsevier Amsterdam.

Vickerman, K., 1992. The diversity and ecological significance of Protozoa. *Biodiv.Conser.,* **1**, 334-342.

Vinogradov, A.P., 1953. The elementary chemical composition of marine organisms. Sears Found. *Mar. Res.*, Mem. **2**, Yale University, 647p.

Walker, G. and Jacob, E., 1798. *In*: **Essays on The Microscope**. F. Kanmacher, 2nd ed., **London**, 629-645.

Walker, G. 1784. Testacea Minuta Rariora Nuperrime Detecta in Arena Litoris Sandvicensis. G. *Boys,* London, 25p.

Wallich, G.C. 1877. Observation on the Coccosphere, *Annals Mag. Nat. Hist.,* sr. **4**(19), 342-350.

Walton, W. R., 1952. Techniques for recognition of living Foraminifera. *Contr. Cush. Found. Foram. Res.*, **3**, 56-60.

Whatley, R.C., Cook, P.C.B. and Warne, M.T. 1996. The ostracoda form lee point on shoal bay, Northern Australia, pt. 2- podocopina (Bairdacea and Cypridacea), *Rev. Espanola Micropal.,* XXVII, **1**, 5-41.

Wilbur, K.M. and Watabe, N., 1963. Experimental studies on calcification of the alga *Coccolithus huxley, Annals, N. r. Acad., Sci.,* **109**, 82-112.

Williamson, 1851. On the minute structure of the calcareous shells of some Recent species of Foraminifera, *Microsc. Soc. Lon.,* Trans., **2**, 159-178.

Wise S. W. And Kelts, K.R., 1972. Inferred diagenetic history of a weakly silicified deep sea chalk, *Trans. Gulf Coast Assoc. Geol. Socs.,* **22**, 1-77.

Wittaker, R.H., 1969. New concepts of kingdom of organisms. *Sci.,* **163**, 150-160.

Wood, 1963. Wall structure of Foraminifera in polarized light. *Micropal.,* **9**(4), 432.

Wood, A., 1949. The structure or wall of the test in the Foraminifera, its value in classification, *Geol. Soc. Lon., Q.J.,* **104**, 229-255.

Systematic Index

Subject Index

D

U

V

W

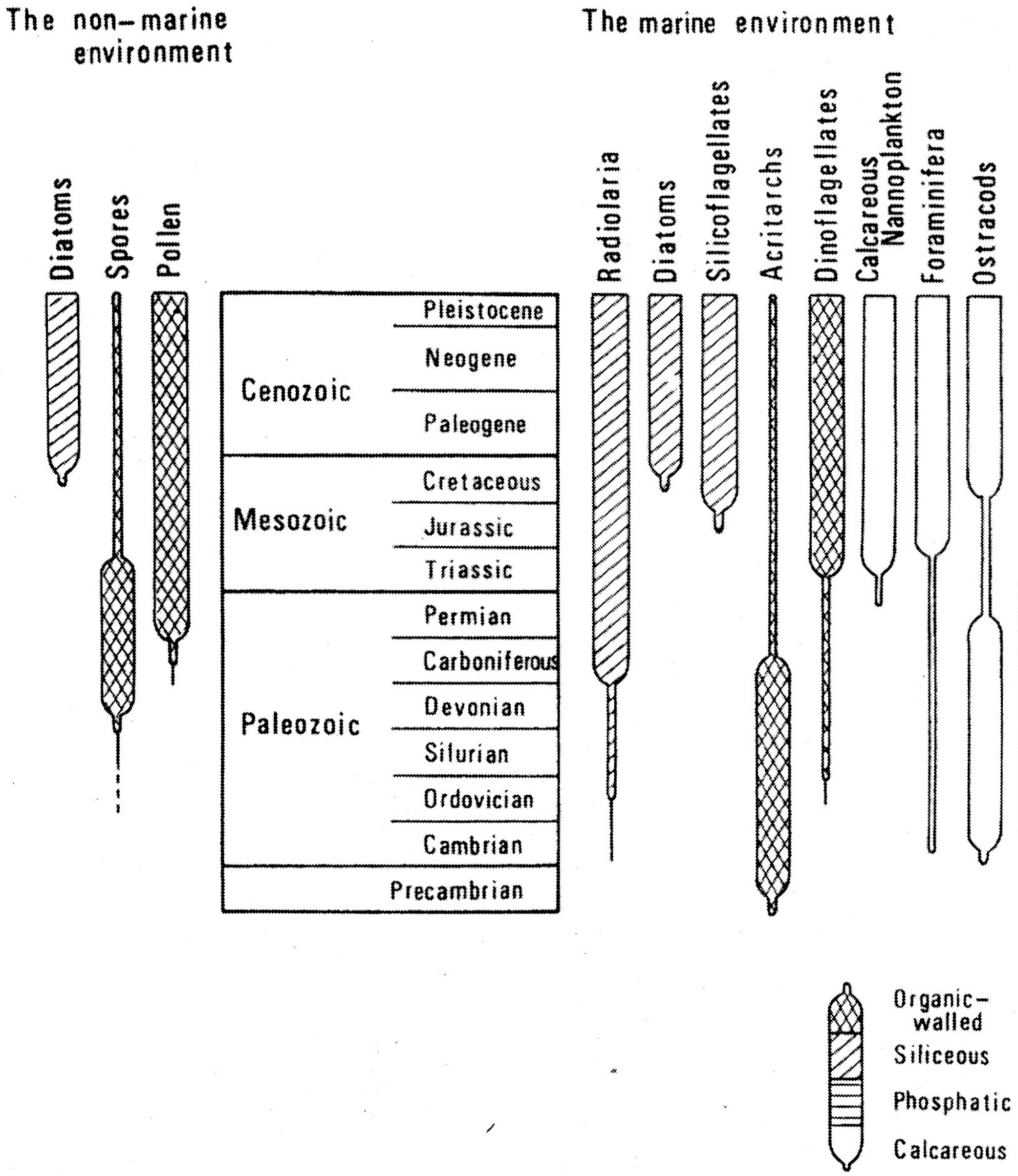

Range of selected microfossil groups with their living environment

Other Related Books

Global Groundwater Resources and Management

Prof. B.S. Paliwal (Editor)

The book on "Global Groundwater Resources and Management" has been aimed at growing recognition of depleting world groundwater resources, their contamination, future requirement of the fast growing human population of the world and urgent need of a better groundwater resources management system. The issues are quite complex but now they have crossed all the political boundaries and have become a common cause. The biggest challenge before the scientific community is to integrate valuable scientific advances and technological progress to solve the issue based on holistic and effective sustainable groundwater management. Each paper has been reviewed by at least two internationally recognized subject experts. In many cases as many as 5 or 6 subject experts have been contacted for the review work for a single paper because a much delayed response from the first two reviewers. Fortunately, all of them have sent their review comments simultaneously but at a later date.

The book is broadly divided into five sections dealing with key aspects of global groundwater resources and management

1. *Groundwater Resources and Management*
2. *Hydogeological Conditions, Groundwater Assessment and Modeling*
3. *Hydrogeochemistry and Contamination of Groundwater Resources*
4. *Exploitation of Groundwater and its Recharge*
5. *Hazardous Groundwater Conditions.*

ISBN: 978-81-7233-619-6, 2010, 528 pp., Hardback, ₹ 3700

Geology of Rajasthan – Northwest Indian Shield

A.B. Roy & S.R. Jakhar

The book will help the future earth scientists to identify potential horizons of future research work not only to enhance our knowledge but also to provide clues to finding new mineral deposits in this paradise for geologists.

Contents

1. Physical Geology; **2.** Geophysics of the Crust of Rajasthan; **3.** Geological Framework; **4.** Basement Rocks-age, Evolution and Reconstitution; **5.** Palaeoproterozoic : Aravalli Fold Belts; **6.** Mesoproterozoic Delhi Fold Belt; **7.** Meso-neoproterozoic Cratonic Basins : Vindhyan Supergroup; **8.** Neoproterozoic Geology of Western Rajasthan; **9.** Phanerozoic Stratigraphy (Palaeozoic To Tertiary); **10.** Mesozoic-cenozoic Magmatsm in Rajasthan; **11.** Quaternary Geology and Neotectonism; References

ISBN: 81-7233-304-8, 2002, 421 pp., Hardback, ₹ 1900

Other Related Books

Economic Mineralization

Prof. K.L. Shrivastava

Economic Mineralization - the volume sets out to present various aspects of a very broad details of a narrow field of economic mineralization at a time when the competitively growing global economy and the pressing needs of the society are compelling economic geology to grow and pile of data is accumulating and opinions changing very rapidly.

The volume incorporates papers, a resultant of information explosion and electrifying conceptual revolution in economic geology, describing the new and exciting results and timely reviews integrating and immense amount of knowledge in the field of geology, exploration, mining, environment, economics, geophysics and geochemistry that has bearing on economic mineralization.

The book imbibes sections on crustal evolution and economic mineralization, economic mineralization of igneous application, economic mineralization of sedimentary affiliation, prospecting and exploration and mining, economics and environments. In all the five sections current concepts, problems and probable trends of future research are highlighted. This book will be an invaluable everlasting reference for both industry and academia specializing in economic mineralization and for those who need updated information and current research in the field. It will be equally useful for advance level geology and mining students and research scholars throughout the world.

Highlights

1. The topic chosen by the authors are of very high standard and also need of the time.
2. All the authors (ninety) are distinctively internationally recognized experts of the discipline.
3. The volume is also unique because reviewing of papers in such volumes is not a common practice. All the fifty papers in the volume were reviewed by the editor and invariably scrutinized and approved by atleast two external referees.
4. All the reviewers (one hundred and seven) are very high profile experts of the field.
5. Every section of the volume is introduced by high quality editorial comments as preamble.
6. A number of coloured photographs, lend emphasis to the narrative of the papers.
7. All papers carry a large number of references that can be consulted frequently and extensively.

ISBN : 978-81-7233-572-4, 2009, 545 pp., Hardback, ₹ 3600

Concise Glossary of Geology

S.M. Mathur

This small volume serves to fill this gap, and covers all the terms that may ordinarily be required. The definitions given are up-to-date and where applicable follow those of the International Union of Geological Sciences.

ISBN: 81-7233-461-3, 2007, Hardback, 156 pp., ₹ 225

Remote Sensing

Principles and Applications

A.N. Patel & Surendra Singh

This book deals with the basic principles of remote sensing such as definition and types of remote sensing, electro-magnetic radiation & its interaction with matter, types of sensors and platforms, data products and visual & digital analysis and interpretation techniques.

ISBN: 81-7233-214-9, 2007, Hardback, 220 pp., ₹ 325